Akitoshi Hanazawa, Tsutom Miki, and Keiichi Horio (Eds.)

Brain-Inspired Information Technology

Studies in Computational Intelligence, Volume 266

Editor-in-Chief
Prof. Janusz Kacprzyk
Systems Research Institute
Polish Academy of Sciences
ul. Newelska 6
01-447 Warsaw
Poland
E-mail: kacprzyk@ibspan.waw.pl

Further volumes of this series can be found on our homepage: springer.com

Vol. 245. Oleg Okun and Giorgio Valentini (Eds.)
Applications of Supervised and Unsupervised Ensemble Methods, 2009
ISBN 978-3-642-03998-0

Vol. 246. Thanasis Daradoumis, Santi Caballé, Joan Manuel Marquès, and Fatos Xhafa (Eds.)
Intelligent Collaborative e-Learning Systems and Applications, 2009
ISBN 978-3-642-04000-9

Vol. 247. Monica Bianchini, Marco Maggini, Franco Scarselli, and Lakhmi C. Jain (Eds.)
Innovations in Neural Information Paradigms and Applications, 2009
ISBN 978-3-642-04002-3

Vol. 248. Chee Peng Lim, Lakhmi C. Jain, and Satchidananda Dehuri (Eds.)
Innovations in Swarm Intelligence, 2009
ISBN 978-3-642-04224-9

Vol. 249. Wesam Ashour Barbakh, Ying Wu, and Colin Fyfe
Non-Standard Parameter Adaptation for Exploratory Data Analysis, 2009
ISBN 978-3-642-04004-7

Vol. 250. Raymond Chiong and Sandeep Dhakal (Eds.)
Natural Intelligence for Scheduling, Planning and Packing Problems, 2009
ISBN 978-3-642-04038-2

Vol. 251. Zbigniew W. Ras and William Ribarsky (Eds.)
Advances in Information and Intelligent Systems, 2009
ISBN 978-3-642-04140-2

Vol. 252. Ngoc Thanh Nguyen and Edward Szczerbicki (Eds.)
Intelligent Systems for Knowledge Management, 2009
ISBN 978-3-642-04169-3

Vol. 253. Roger Lee and Naohiro Ishii (Eds.)
Software Engineering Research, Management and Applications 2009, 2009
ISBN 978-3-642-05440-2

Vol. 254. Kyandoghere Kyamakya, Wolfgang A. Halang, Herwig Unger, Jean Chamberlain Chedjou, Nikolai F. Rulkov, and Zhong Li (Eds.)
Recent Advances in Nonlinear Dynamics and Synchronization, 2009
ISBN 978-3-642-04226-3

Vol. 255. Catarina Silva and Bernardete Ribeiro
Inductive Inference for Large Scale Text Classification, 2009
ISBN 978-3-642-04532-5

Vol. 256. Patricia Melin, Janusz Kacprzyk, and Witold Pedrycz (Eds.)
Bio-inspired Hybrid Intelligent Systems for Image Analysis and Pattern Recognition, 2009
ISBN 978-3-642-04515-8

Vol. 257. Oscar Castillo, Witold Pedrycz, and Janusz Kacprzyk (Eds.)
Evolutionary Design of Intelligent Systems in Modeling, Simulation and Control, 2009
ISBN 978-3-642-04513-4

Vol. 258. Leonardo Franco, David A. Elizondo, and José M. Jerez (Eds.)
Constructive Neural Networks, 2009
ISBN 978-3-642-04511-0

Vol. 259. Kasthurirangan Gopalakrishnan, Halil Ceylan, and Nii O. Attoh-Okine (Eds.)
Intelligent and Soft Computing in Infrastructure Systems Engineering, 2009
ISBN 978-3-642-04585-1

Vol. 260. Edward Szczerbicki and Ngoc Thanh Nguyen (Eds.)
Smart Information and Knowledge Management, 2009
ISBN 978-3-642-04583-7

Vol. 261. Nadia Nedjah, Leandro dos Santos Coelho, and Luiza de Macedo de Mourelle (Eds.)
Multi-Objective Swarm Intelligent Systems, 2009
ISBN 978-3-642-05164-7

Vol. 262. Jacek Koronacki, Zbigniew W. Ras, Slawomir T. Wierzchon, and Janusz Kacprzyk (Eds.)
Advances in Machine Learning I, 2009
ISBN 978-3-642-05176-0

Vol. 263. Jacek Koronacki, Zbigniew W. Ras, Slawomir T. Wierzchon, and Janusz Kacprzyk (Eds.)
Advances in Machine Learning II, 2009
ISBN 978-3-642-05178-4

Vol. 264. Olivier Sigaud and Jan Peters (Eds.)
From Motor Learning to Interaction Learning in Robots, 2009
ISBN 978-3-642-05180-7

Vol. 265. Zbigniew W. Ras and Li-Shiang Tsay (Eds.)
Advances in Intelligent Information Systems, 2009
ISBN 978-3-642-05182-1

Vol. 266. Akitoshi Hanazawa, Tsutom Miki, and Keiichi Horio (Eds.)
Brain-Inspired Information Technology, 2009
ISBN 978-3-642-04024-5

Akitoshi Hanazawa, Tsutom Miki, and
Keiichi Horio (Eds.)

Brain-Inspired Information Technology

Springer

Prof. Dr. Akitoshi Hanazawa
Department of Brain Science and Engineering
Graduate School of Life Science and System
Engineering, Kyushu Institute of Technology
Hibikino 2-4, Wakamatsu
Kitakyushu, Fukuoka 808-0196
Japan
E-mail: hanazawa@brain.kyutech.ac.jp

Prof. Dr. Keiichi Horio
Department of Brain Science and Engineering
Graduate School of Life Science and System
Engineering, Kyushu Institute of Technology
Hibikino 2-4, Wakamatsu
Kitakyushu, Fukuoka 808-0196
Japan

Prof. Dr. Tsutom Miki
Department of Brain Science and Engineering
Graduate School of Life Science and System
Engineering, Kyushu Institute of Technology
Hibikino 2-4, Wakamatsu
Kitakyushu, Fukuoka 808-0196
Japan

ISBN 978-3-642-04024-5 e-ISBN 978-3-642-04025-2

DOI 10.1007/978-3-642-04025-2

Studies in Computational Intelligence ISSN 1860-949X

Library of Congress Control Number: 2009940421

Typeset & Cover Design: Scientific Publishing Services Pvt. Ltd., Chennai, India.

Printed in acid-free paper

9 8 7 6 5 4 3 2 1

springer.com

Preface

Present information systems have huge computational power and are growing day by day. Although their applications distribute various fields, what they can do is rather complimentary what we human can do. Computers can execute complicated calculations or can pick up a literature from a huge database, which we human cannot do. They, however, cannot recognize what a human is doing in a movie or cannot understand what he is talking about, which we human can easily perform. This difference between computer systems and the human brain is due to the difference in algorithms employed for various functions, and also due to the difference in hardware. Our brain works in parallel, that is, executes a huge number of calculations at the same time. Thus, to equip computers with such human like functions, brain-like algorithms and devices are necessary.

One more sticking difference between computer systems and the human brain is in their creation processes. For artificial information systems, we must give instructions how they should work about all possible situations. They cannot do anything in unexpected situations. On the other hand, we humans acquire various behaviors based not only on genetic information but also on post-natal learning. Especially, post-natal learning enables us to behave correctly in genetically unexpected situations, e.g. driving a car or reading/writing letters. And also, the learning can correct mistakes. A behavior that does not fit to a situation changes into another better behavior. Moreover, most animals have the ability of generalization, guessing a correct behavior for an unexperienced situation based on learned relationships between behavior and situation. Rigidity is one of very important characteristics of computer systems, because we need accurate results of calculations or text searches. Adding human-like flexibility by learning or self-organization algorithms makes present computer systems more powerful and usable.

To realize brain-like information technology, we have still not obtained enough knowledge about how our brain works. Brain science has been producing huge amount of data about our brain. However, many keys to understand our brain function, such as mechanisms of memory, learning, recognition, and etc., are still in a fog of complexity. Some researches intend to apply the data on the development of information system, and others outputs fragments of knowledge useful for constructing information technology, which diffuses and drifts in the sea of knowledge about our brain. Thus technological development of brain-like information system and scientific brain research are necessary to interact to pick up useful fragments of the knowledge or to inspire each other.

Contents

Mission Capable Autonomous Control Systems in the Oceans, in the Air and in Space

Sandor M. Veres

School of Engineering Sciences, University of Southampton, Highfield, Southampton, SO17 1BJ, UK
s.m.veres@soton.ac.uk

Abstract. The problem of optimal degree of autonomy and minimal complexity is considered for mission capable systems in spacecrafts and unmanned aeronautical systems. Levels of autonomy and system requirements are discussed. Simple and goal oriented system architectures are summarized and the prospect of plug-and-play systems is outlined. The engineering compromise to be made between degree of autonomy and verifiability of multi-agent systems is pointed out. Three main challenges are highlighted to aid progress of future development of verifiable autonomous vehicles.

1 Introduction

Autonomous system have been inspired and continue to be inspired by biological systems of nature. The concept and use of feedback for telephony repeaters (Bell Labs 1930s) and later control theory (Bode) and cybernetics (Wiener) introduced the first kind of autonomy where the engineered system reliably regulated to a reference on its own, i.e. without human supervision. Some may debate whether a robust feedback loop is autonomous control but it nevertheless has the property of independent operation. Later autonomous control included the ability of the system to set its reference signals and its own parameters for control depending on the circumstances and its goals. Higher levels of autonomy could then be defined by the high level of command that was possible to issue that the autonomous system executed on its own. For instance the command to an autonomous aerial vehicle (AAV) "go to position A and take a photograph" requires the system to plan its route under constraints and then execute it while coping with instantaneous obstacles and problems on the way and finally coordinating its motion to take a quality photo and plan and execute its return. This command still shows a much lower level of autonomy than telling a similar command to an autonomous ground vehicle (AGV) in urban environment such as "fetch Clive James from the Nuffield theatre and bring him here". The higher level comes from the much more complex set of constraints in an urban-human environment

A. Hanazawa et al. (Eds.): Brain-Inspired Info. Technology, SCI 266, pp. 1–10.
springerlink.com

including, driving on a busy road, finding parking place, parking, calling Clive James on his mobile where the AGV ("me") is and asking him to sit in, etc. are in themselves difficult to plan, coordinate and execute.

At high altitudes in the air, midwater in the oceans and space and in orbit around the earth the complexity of the environment can be less than in an urban and human environment or on complex wild terrain on the earth surface. It is therefore a reasonably good start to design and build autonomous systems for aircrafts (AAV), submarines (AUV) and spacecraft (S/C).

2 Autonomous AUVs, AAVs and S/C at Southampton

Various types of autonomous systems are developed at the University of Southampton. Enhancing their reliability by new intelligent agent techniques is subject to intense current research.

Three past generations of the Autosub I-II-III have been developed at the National Oceanographic Centre at Southampton University [4] for science missions with a maximum ocean depth of about 1600m. The Autosub 6000 is complete redesign that has recently been built and passed a test at a depth of 5400m. The main use of autosubs are arctic ice shelf explorations, deep sea biological and geological sample, topographic surveys, global warming studies, research into climate changes, sea current measurements, chemical analysis of seawater, etc.

Fig. 1. Autosub 6000 of the National Oceanography Centre in Southampton

Autonomous aerial vehicles (AAVs) are currently under development in student group design projects and PhD research projects. The Io UAV [5] is going through its second generation of development for oceanographic use to assist the autosubs in science missions in a multi-agent system.

Fig. 2. The first generation of the Io UAV [5]

Current small aeroplanes in use are at the School of Engineering Sciences (SES) and at NOC:

T60 – experimental NOC plane
T240 – Student project plane at SES
IO-1 – NOC ship support plane
IO-2 – long range NOC ship support plane
AAVx – planes designed and built in GDP projects at SES

Autonomous spacecraft research is based around a nano-satellite autonomous control hardware and software testing facility that enables testing of distributed agent systems with processor hardware, communications hardware and sensors such solid state gyros, accelerometers, camera systems, lidars, magnetometers and actuators such as reaction wheels and control moment gyros. The space certified electric and cold gas thrusters need however be replaced by ones that can operate in the atmosphere, hence either propellers or pressurised air thrusters can be used.

The Spacecraft Autonomy Centre at the SES provides a complementary set of testing facilities:

- Sensors-actuators-processors-communications (SAPC) testing except thrusters. This testing is done with control software suitable on the 5DOF satellites in gravitational field on the ground and not in space.
- Hardware-in-the-loop (HILT) testing where nano-satellite frames are fixed and satellite dynamics in space, sensors and actuators are all simulated with realistic noise on all sensor measurements. In this test the on-board processor, software and communication hardware is used as in space.

The hardware testing facility is being made suitable to test finished satellites with the use of pressurised air thrusters instead of the space qualified ones. Additional benefits are for the development of autonomous multi-agent systems ins space.

Fig. 3. 5DOF (two translational and 3 rotational DOF) satellite frames on which autonomous control systems can be tested

3 Levels of Autonomy

This section is about basic classification of practical autonomy levels for the autonomous vehicles in consideration. Humanoid robots, pets and ground vehicles may have possibly higher levels for autonomy but we restrict ourselves to minimal practical requirements that can also lend themselves to verification that is often missing for more complex autonomy levels.

Table 1. Classification of autonomy levels for autonomous vehicles

None	Remotely controlled control surfaces and thrust: "model aircraft with remote control unit" - no autonomy
Low	The autonomous vehicle follows a sequence of waypoints, carries out scheduled task execution such as photography, performs sensor fusion and some visual analysis – low level of autonomy
Medium	Ability to cope with seriously adverse conditions, performs path and action re-planning and execution to reach geographic targets and persists to successfully perform tasks – medium level of autonomy
High	"Understands" a multilayered model of a constrained environment where high level goals can be interpreted

At a low level of autonomy the vehicle should already plan its path to the next waypoint by taking into account vehicle dynamics, fuel efficiency and obstacles. Also it should plan and timely execute mission related tasks such as photography or collecting samples from the environment.

At medium level of autonomy more complex abilities are required such as decision making under rapidly changing environmental conditions that were not anticipated. Realtime complete re-planning of the mission path or re-planning of task execution should be carried out by the agent system on board. Also a richer set of high level commands that contain conditional statements should be possible to issue to the autonomous vehicle.

At high level of autonomy the instructions to the autonomous vehicle are issued at the level of goal statements and the autonomous system should be able to work out a hierarchical plan system based on multi-resolution modeling of the environment under action constraints. This is the level of behaviour that one would call "intelligent" in the common sense of the word.

3.1 Ingredients of Creating Low Level of Autonomy

The most important points for low level of autonomy are as follows.

- Good navigation system by data fusion of GPS, Inertia Measurement Unit (IMU), depth/altimeter and other sensors typically using extended Kalman or/and neuro-fuzzy filtering
- Robust controller that is suitable for tracking
- Relatively low processing power CPU memory <100KB
- For implementation a multi-agent system is not a necessity

3.2 Ingredients of Creating Medium Level of Autonomy

In addition to the capabilities of low level autonomous systems, the following are also needed to achieve medium level autonomy:

- Onboard re-planning algorithms of waypoints
- Simplest suitable agent architecture, parallel processing
- Reconfiguration algorithms in case of sensor failure
- Multiple processors and large memory

3.3 Ingredients of Creating High Level Autonomy

In case of high level of autonomy a sophisticated but verifiable multi-agent system is needed that has some meta knowledge capabilities.

- Onboard knowledge of missions & operational environment constraints and own capabilities
- Ability to interpret a highly abstract goal given in terms of a language describing the environment.
- Ability to generate sub-goals and sub-plans to achieve higher level goal
- Realtime modelling and ability to re-generate plans as conditions change fast
- Large (multi-)processor capacity and multi-agent system
- Onboard knowledge of missions & operational environment constraints and own capabilities
- Ability to interpret a highly abstract goal given

4 Algorithmic Architectures

Event the low levels of autonomy need good navigation based on data fusion to locate the vehicle relative to its desired locations.

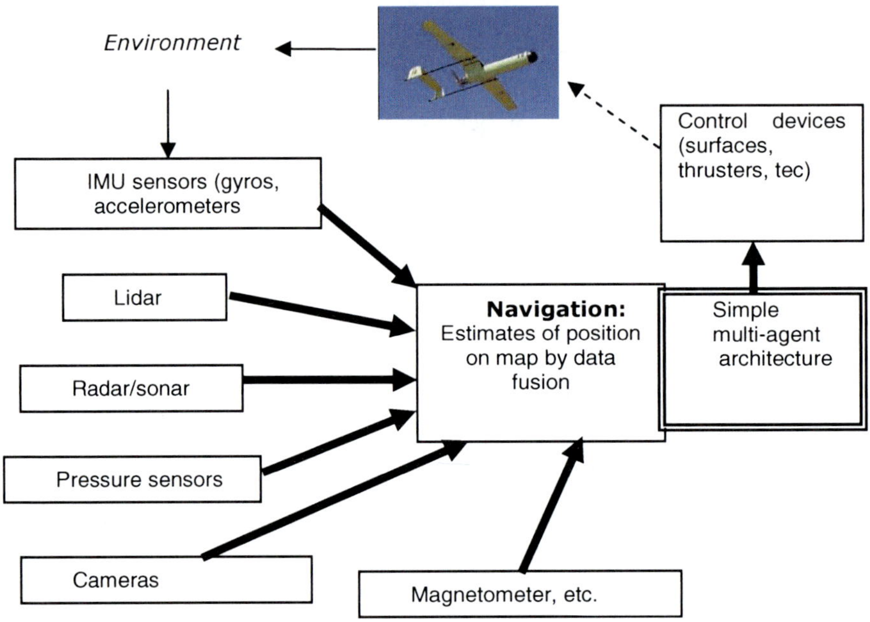

Fig. 4. Navigation system components

Typical navigation systems use various versions of extended Kalman filtering and fast implementations of image processing, possibly massively parallel computations for visual target localization and recognition from camera pictures. Frequency domain based feature recognition of signals is commonplace in navigation systems.

- At the lowest level of agent architectures are the *reactive (behavioural) agents* (low level): the sensed environmental situation determines the behaviour mode, intelligence emerges from agents-nature interaction
- At medium level there are the *deliberative architectures* that use some kind of world model, prediction and planning to choose from version of possible actions (see illustration in Fig. 5).
- At high level well know architectures are the *bliefs, desires, intentions* (BDI) deliberative architectures that use goal setting, planning, intentions and desire (~sub-goal) generation (see illustration in Fig. 6). Alternatives are suitable *multilayered architectures* for high levels of autonomy (see illustration in Fig. 7).

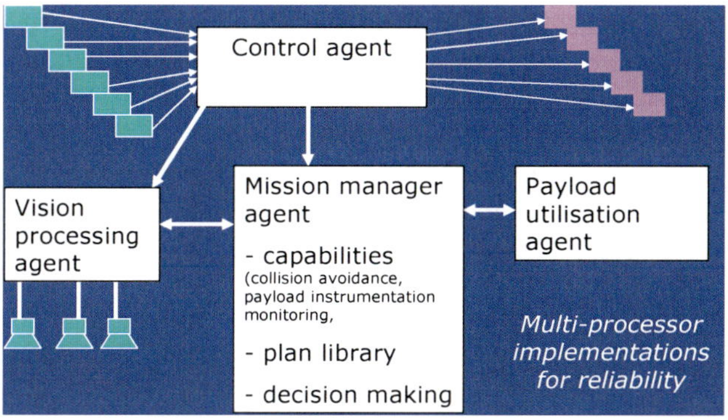

Fig. 5. Simple multi-agent architectures for medium level of autonomy

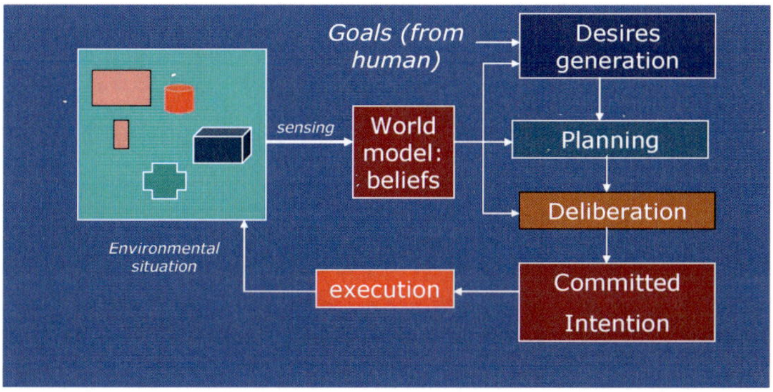

Fig. 6. The beliefs desires intentions architecture for high levels of autonomy

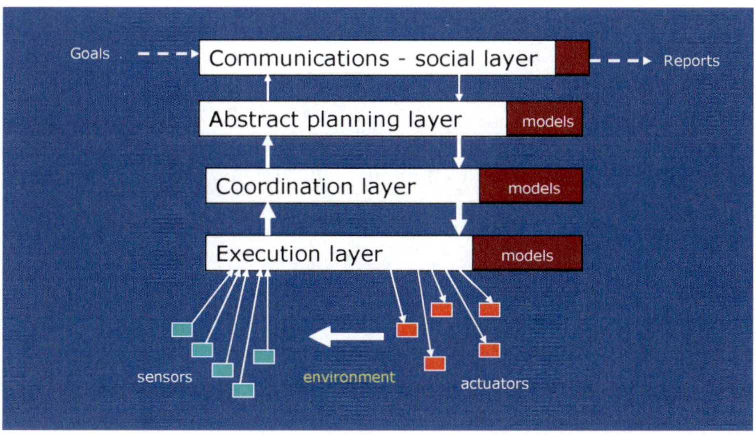

Fig. 7. The multilayered architecture for high levels of autonomy

Multi-layered architectures normally have communications at the top of their layers. Information flows sensors to control loops in the execution layers. From there abstracted information is passed to the coordination layer and abstract planning layers and finally to communications that decides whether interaction with other agents (including humans) is necessary. Once decisions have been made information flows downwards to work out the details of actions.

The above three level low-medium-high classification does not account for humanoid robot intelligence (motivations, emotions, playfulness, comprehensive learning ability, social behaviour, shared knowledge development, etc. are not developed here). The purpose of AUVs, AAVs and S/C is to be good reliable servants to achieve human goals.

The behaviour of these vehicle systems must be reliable and verifiable for functionality and safety even in the case of hardware malfunction. Hardware and agent system verification is needed that leads to the application of model checking of multi-agent systems. At Southampton and Imperial College London the author coordinates an EPSRC project (£720k) on verification of AUV engineering/agent systems for mission reliability.

5 Challenges Ahead

There are many problems that engineers face while developing autonomous vehicle systems. Obvious challenges are in relation to human-system communications. The problem of operator performance and intelligent aiding in unmanned aerial vehicle scheduling and choice of varying levels of autonomy levels has been discussed in [3]. Learning to control complex and variable dynamics is also a challenge. In [2] a self-organizing map (SOM)-based local linear modeling scheme of an autonomous aerial vehicle (AAV) was developed to design a set of inverse controllers. The SOM selects the operating regime depending only on the embedded output space information and each local linear model is associated with a linear controller.

We will pick from remaining challenges three that are crucial for practical progress and efficiency of designs.

Challenge 1: Reliable *navigation systems using data fusion methods* (plug and play (PnP) data fusion for modelling the environment. Algorithms for new sensor arrays and dedicated hardware architectures for signal processing of visual and audio perception.

Challenge 2: High level *agent architectures that are verifiable by design.* Use of agent definition formalism that easily translates into model checking. Modularisation of capabilities and behaviour modes to enable hierarchical model checking.

Challenge 3: *Compatible agent oriented programming tools* and human-machine communication systems that are easy to use without too much training as it is today.

5.1 Navigation Problem Examples

(a) Determine the realtime position of the AUV based on range only measurements. A difficulty is that the initial position of the AUV is poorly known: before it starts its mission path it floats and drifts for a while not far from the ship and submerged that makes GPS fix not applicable. Joint estimation current and initial position is possible if inertial measurement system and range only measurements on the ship are available for a data fusion algorithm.

(b) High precision relative position/attitude estimation for a cluster of spacecrafts. The difficulty is that there are no other reference objects nearby. Possible solution is that each satellite has some active or passive vision system and the resulting information is shared and "fused" realtime [6].

(c) Camera vision based realtime navigation of AAVs is to be made faster. Extended Kalman filter based realtime navigation works assuming objects on the ground are recognised in timely manner [1]. Terrain aided navigation system that has the capability for online map building, and simultaneously utilizing the generated map to bound the errors in the navigation solution. Recognition of more complex objects in a cluttered environment can be too slow. Parallel processing of 3D feature recognition under varying light conditions is very much needed through dedicated hardware.

5.2 Creating Agent Architectures for High Level of Autonomy

Software features needed are:

- Easy to use multi-agent system prototyping environment that has single console for multi-agent system built/launch/supervise
- It should allow integration of high level sensor signal processing and vision programming code
- Should be possible to create prototypes of reactive/behavioral/BDI and multilayered agents with high level code at ease.
- Agent prototypes should be possible to transfer to C++/C#/ Visual C/ Robotics Studio implementations
- Should permit fast TCP/IP communications between agents
- Should provide free access to some valuable code by user community

5.3 Compatible Agent Oriented Programming for Engineers

Different programming platforms are in use – but the programming language is not the major issue. The major issue is that the software produced cannot easily be integrated – plug and play & self assembly of agents is needed from components.

How to achieve wide ranging compatibility of software? Ontology based programming which can be viewed as an extension of object oriented programming can be used. There is OWL but it is not directly applicable in agent programming.

A possible solution is the machine ontology language (MOL) that defines conceptual structures in a modelling dimension of a functional group. A further desirable feature for development engineers at companies and enthusiast is natural language based programming using the ontology classes. Initial steps have been made in this direction [8].

6 Conclusions

This paper has first outlined the practical work carried out at the university of Southampton. Practical work and reliability requires the least complexity from autonomous systems, hence a limited scope of autonomy has been outlined. The most relevant practical agent architectures have been reviewed. Finally the discussion has been brought to its logical conclusion on the challenges ahead that includes fast enough navigation systems, agent programming tools for engineers and compatibility of software development via natural language programming and the machine ontology language.

References

1. Kim, J., Sukkarieh, S.: Autonomous Airborne Navigation in Unknown Terrain Environments. IEEE Trans. Aero. Elec. Systems 40, 1031–1044 (2004)
2. Cho, J., Principe, J.C., Erdogmus, D., Motter, M.A.: Modeling and Inverse Controller Design for an Unmanned Aerial Vehicle Based on the Self-Organizing Map. IEEE Trans. Neural Netw. 17, 445–460 (2006)
3. Cummings, M.L., Brzezinski, A.S., Lee, J.D.: Operator Performance and Intelligent Aiding in Unmanned Aerial Vehicle Scheduling. IEEE Intelligent Systems Magazine, 52–57 (March/April 2007)
4. Griffith, G. (ed.): Technology and Applications of Underwater Vehicles. Taylor and Francis, Abington (2002)
5. Waugh, E., Mowlem, M., Veres, S.M.: Aerodynamic optimization of an unmanned aerial vehicle for oceanographic applications (2007), `http://www.soton.ac.uk/engd/events/2007_conference/posters/Waugh_Aerodynamic_optimisation.ppt`
6. Lincoln, N.K., Veres, S.M.: Components of a vision assisted constrained autonomous satellite formation flying control system. Int. J. of Adaptive Control and Signal Processing 21, 237–264 (2007)
7. Veres, S.M.: sEnglish and CAT, `http://www.sysbrain.org`
8. Veres, S.M.: Editorial to the special issue on autonomous adaptive control of vehicle systems. Int. J. of Adaptive Control and Signal Processing 21, 93–94 (2007)

Three Paths to Relevance

Samuel Kaski

Adaptive Informatics Research Centre and Helsinki Institute for Information Technology, Helsinki University of Technology. P.O. Box 5400, FI-02015 TKK, Finland
samuel.kaski@tkk.fi

Abstract. The problem of distinguishing between relevant and irrelevant variation is shared by natural and artificial data analysis systems. The problem is especially hard when the dimensionality of the data is high and yet inferences need to be made based on only few samples, and when the models are flexible machine learning models. I will discuss three learning strategies for coping with the problem, and applications in brain imaging, bioinformatics, and information retrieval.

1 Introduction

"Natural signals," signals measured with imperfect sensors from systems that cannot be fully controlled, necessarily contain irrelevant variation, that is, variation the receiver is not interested in. The signals may range from scientific measurements to more abstract "signals" in the newsfeeds of the Internet, and the applications may range from brain imaging to information retrieval.

Tackling the irrelevant variation is hard in the common setting where (i) it is complex, and (ii) there are only few samples available for learning the task. The latter part of the problem has been called the large p small n problem in the context of bioinformatics, where it is hard to decide which of the genes (dimensions p) are differentially expressed in a treatment, when the number of measurement samples n is small.

A traditional way of coping with irrelevant variation is to assign a model for it. The irrelevant part is often called noise, and some simple assumptions are typically made about its distribution. For instance, it may be assumed that the measurements are an additive mixture of normally distributed noise ε and relevant signal, $x = f(s, \theta) + \varepsilon$, where the function f describes how the "latent sources" s are transformed into the observations x. The model is then fitted to the data by optimizing the parameters θ.

This process works if the distributional assumptions about the noise have been reasonable and if there is enough data. If it is known that the noise is complex but its structure is not known, we do not have a suitable model for ε and we will need to do something else. Such irrelevant structured variation could alternatively be called a nuisance signal; it is noise in the sense of being irrelevant to the user or processing task.

A. Hanazawa et al. (Eds.): Brain-Inspired Info. Technology, SCI 266, pp. 11–13.
springerlink.com

2 Learning Metrics

The problem of distinguishing between relevant and irrelevant variation is particularly hard for *unsupervised learning*, where the task is to find statistical structures or dependencies within a data set. In fact, the task is impossible unless more is known of the data or the task; this problem has been called the *garbage in—garbage out* problem: if an unsupervised learning system is fed with uninteresting data it will produce uninteresting results.

The situation changes if *auxiliary data c* are available during learning, such that the auxiliary data are known to be relevant or interesting. This works even for unsupervised learning. We assume that the data comes in pairs *(c,**x**)* where the *c* could be classes of the samples ***x*** in the task of studying what the classes are like and how they differ. We know, based on the setup, that those aspects of ***x*** that are related to variation in *c* are relevant and the rest are irrelevant. We have developed methods for *learning metrics* to the ***x***–data spaces using the paired samples *(c,**x**)* as the learning data, such that the metrics discard the irrelevant variation. The metrics can be used in many standard data analysis methods to mine or explore the data ***x*** [2,6]. Main applications so far have been in gene expression analysis and finance.

The task for which the metrics are learned could be called *supervised unsupervised learning* or *supervised mining*; the results are relevant for the classes *c*.

3 Learning of Mutual Dependencies in Data Fusion

For co-occurring data *(**x**,**y**)* another kind of approach is possible, assuming the relationship between ***x*** and ***y*** is interesting but neither of the samples ***x*** or ***y*** is known to be more relevant than the other as in learning of the metrics above. The sample pairs can be decomposed into components that are specific to each sample, and components that are shared. A classical method for finding correlated components is the Canonical Correlation Analysis (CCA). We have extended it to models of local components to remove the assumptions of global linearity and normality underlying CCA, and to decompose the data to shared and data-specific components [4]. A Bayesian treatment made it possible to tackle overfitting to small numbers of samples, which is a hard problem for CCA.

The task of data fusion to search for mutual dependencies between data sources, or mining of what is shared, could be called mutual dependency mining or exploration. Relevance is here inferred from co-occurring data. Applications have so far been in defining yeast stress response [5], searching for dependencies in gene expression between man and mouse [3], and in fMRI imaging of the brain [7].

4 Learning from Relevant Tasks

The third alternative to combat the large *p*, small *n*–problem is to get more data. For instance, in molecular biology there are large databases of gene expression data available. The problem is that typically only a part of the data is relevant.

We have recently introduced a variant of multi-task learning for this problem [1]. Assume that there are several different data sets or "tasks" available, each containing some relevant data samples. We model each set as a mixture of relevant and irrelevant data, such that the irrelevant data become "explained away" by their own flexible model. The model for relevant data is shared by all the tasks. It becomes learned with all the relevant data, in effect increasing the amount of learning data for the interesting task.

Acknowledgements

This research was supported by the Academy of Finland and the PASCAL EU FP6 Network of Excellence.

References

1. Kaski, S., Peltonen, J.: Learning from relevant tasks only. In: Kok, J.N., Koronacki, J., Lopez de Mantaras, R., Matwin, S., Mladenič, D., Skowron, A. (eds.) ECML 2007. LNCS (LNAI), vol. 4701, pp. 608–615. Springer, Heidelberg (2007)
2. Kaski, S., Sinkkonen, J., Peltonen, J.: Bankruptcy analysis with self-organizing maps in learning metrics. IEEE Transactions on Neural Networks 12, 936–947 (2001)
3. Kaski, S., Nikkilä, J., Sinkkonen, J., Lahti, L., Knuuttila, J., Roos, C.: Associative clustering for exploring dependencies between functional genomics data sets. IEEE/ACM Transactions on Computational Biology and Bioinformatics 2, 203–216 (2005)
4. Klami, A., Kaski, S.: Local dependent components. In: Ghahramani, Z. (ed.) Proceedings of ICML 2007, the 24th International Conference on Machine Learning, Omnipress (2007)
5. Nikkilä, J., Roos, C., Savia, E., Kaski, S.: Explorative modeling of yeast stress response and its regulation with gCCA and associative clustering. International Journal of Neural Systems 15, 237–246 (2005)
6. Peltonen, J., Klami, A., Kaski, S.: Improved learning of Riemannian metrics for exploratory analysis. Neural Networks 17, 1087–1100 (2004)
7. Ylipaavalniemi, J., Savia, E., Vigário, R., Kaski, S.: Functional elements and networks in fMRI. In: Proceedings of ESANN 2007, the 15th European Symposium on Artificial Neural Networks, d-side, Belgium (2007)

Brain-Like Evolving Spiking Neural Networks for Multimodal Information Processing

Simei Gomes Wysoski, Lubica Benuskova, and Nikola Kasabov

Department Knowledge Engineering and Discovery Research Institute,
Auckland University of Technology, AUT Tech Park, 581-585 Great South Road,
Ronald Trotter House, Auckland, New Zealand
nkasabov@aut.ac.nz

Abstract. Despite of much evidence suggesting how and where sensory information converge in the human brain, the neural mechanisms of interaction among modalities at the level of neuronal cells and ensembles are still not well understood. The chapter explores emulation of multimodal information processing in a brain-like manner through evolving spiking neural network (ESNN) architectures that use several multimodal characteristics of the biological brains, e.g., multisensory neurons, crossmodal connections, capacity of lifelong adaptation and evolution, adaptive pattern recognition. Illustration is given on audiovisual ESNN for the person authentication problem. Preliminary results show that the integrated system can improve the accuracy in many operation points as well as it enables a range of multi-criteria optimizations.

1 Introduction

There is strong experimental evidence that integration of sensory information occurs in the mammalian brain and a lot is known about the location in the brain where different modalities converge [6, 28, 31, 32]. In simple terms, the integration occurs in *supramodal* areas that contain neurons sensitive to more than one modality, i.e., neurons that process different types of information. Further, *crossmodal* coupling, which is related to the direct influence of one modality to areas that intrinsically belong to other modalities, is another integrative phenomenon noticed in behavioural observations and electrophysiological experiments. This arrangement is illustrated in Figure 1. A good starting point to understand crossmodal learning can be found in [14, 18].

However, studies of neuronal mechanisms that underlie interaction among modalities at the level of single or ensemble of neural cells are still inconclusive. In this direction, computational models of interactions at a neuronal level inspired by perceptual studies can help to shed more light on the modular interdependences, and, in the same manner, the better understanding of these interactions can provide new insights to enhance performance of connectionist algorithms applied to pattern recognition. The latter is the immediate objective of our research.

A. Hanazawa et al. (Eds.): Brain-Inspired Info. Technology, SCI 266, pp. 15–27.
springerlink.com

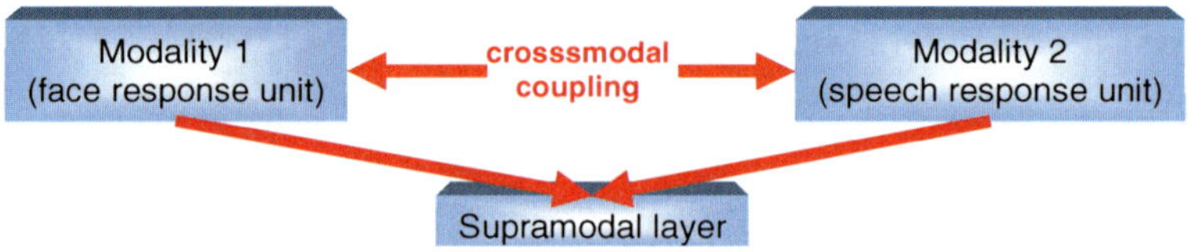

Fig. 1. Sensory integration. Supramodal region and the crossmodal coupling of modalities applied for audiovisual integration.

The integration of modalities for the purpose of pattern recognition is often used in tasks that cannot be solved by a single system or can be facilitated by using more than one source (generally where there is unimodal ambiguity, unimodal lack of data and/or correlation among modes). Many works report significant performance improvement [16, 24, 25, 27] as well as describe that the use of modularity results in systems easy to understand and modify. Added to that, modular approaches are known for contributing against modular damage and for facilitating training and the inclusion of prior knowledge [24].

As example of multi-modal models that use statistical methods of computation (Maximum Likelihood, PCA/ICA, clustering, Gaussian mixture of models), Haller et al. present an audiovisual system to detect anchorperson [15]. The system process modalities separately and the decision fusion is made with AND and OR gates. Park et al. perform audiovisual human authentication using PCA on visual information and cepstral LPC (linear prediction coefficients) to describe auditory features and Hidden Markov Models (HMM) for classification [20]. The integration of modalities is done through an approach that uses fuzzy logic. Another system designed by Ben-Yacoub et al. for audiovisual person authentication is presented in [1], where the multimodal fusion of experts is presented to add robustness and increase performance of authentication process. In this work the individual modalities are processed separately. The face is represented with elastic graph matching (EGM) and features extracted with Gabor responses. Features between two face models are matched with a given distance equation. The speech information is represented with cepstral LPC part and a statistical method to calculate similarity between speakers called arithmetic-harmonic sphericity [2] is used in the text-independent setup. Also a text-dependent system based on HMM is presented. Nonetheless, the main subject of evaluation in [1] is the fusion part, with five different post integration methods being proposed and compared. They are: SVM-based (support vector machine) fusion, minimum cost Bayesian classifier, Fisher linear discriminant, C4.5 classifier, and neural network-based (multi-layer perceptron) classifer. Experiments were carried out with the XM2VTSDB database, which contains synchronized recordings of facial properties as the individual utter digit from 0 to 9 [19]. Brunelli and Falavigna present a system where two classifiers are used on the speech signals and three others classify visual inputs [3]. The results of these individual classifiers are sent to the input of a new integration module based on HyperBF networks [21]. Speech features are MFCC and the corresponding derivatives, and each speaker is represented with a set of vectors based on Vector Quantization (VQ) [23]. A local template matching approach at the pixel level, where particular areas of the face (eyes, nose, mouth) are compared with a previously stored data is used to face authentication. To further improve the performance of the multi-modal systems, there are several methods that propose adaptation of the fusion mechanisms (see [8] for a comprehensive list).

Based on brain-like approaches, Maciokas and Goodman tackle the problem of integrating the visual information of lip movements with the corresponding speech generated by it, with a biologically realistic spiking neural network [17]. 25,000 neurons have been used, placed in 10 columns and several layers. Tonotopic maps fed from Short Time Fourier Transform with a neural architecture that resembles MEL scale filters was used on converting audio signal to spikes. Gabor Filters extracted the lips movements. It was demonstrated that three different sentences were encoded in different spiking patterns and after using Hebbian learning for training, the output spiking patterns were also distinguishable.

Seguier and Mercier also presented a system for integrating lips movements and speech signal to present a one-pass learning with spiking neurons, with performance in favour of the integrated system, mainly under the presence of noise in the audio signal [26]. The system intended to produce real-time results, therefore simple visual features were used and auditory signals were represented with 12 cepstral coefficients. Chevallier et al. present a system based on SNN to be used in a robot able to process audiovisual sensory information in a prey-predator environment [7, 9]. The system is composed of several neural networks (prototype-based incremental classifier), one for each sensorial modality. There is a centralized compartment for data integration, which is implemented as a bidirectional associative memory, and a network (also incremental) used to perform the final classification. The integration of several streams of incoming data is also processed on the fly as soon as the data from different modalities are made available. Further, the bidirectional associative memory implemented with the spiking mechanism enables the simulation of crossmodal interaction.

In this chapter we look a little bit closer at the biological aspects that drive the integration of sensory modalities to present a system that integrates data from different modalities for the purpose of pattern recognition. The processing of individual modalities is implemented using adaptive SNNs. An integration procedure at the neuronal level is presented, which considers crossmodal interrelation among modalities, emulating what has been noticed in several biological experiments. In Section 2 we describe the architecture of the individual systems based on SNNs as well as the integration procedure. Section 3 presents some computational simulations and shows preliminary results when the system is applied to the person authentication problem. A discussion on the main properties of the integrated system and future directions conclude the paper.

2 SNN-Based Multi-modal Pattern Recognition

Our approach of biologically inspired integration of modalities for pattern recognition uses the theory of SNN, where the individual modes and the integration procedure are implemented with spiking neurons. We are using a simplified version of an integrate-and-fire neuron as in [10]. Neurons have a latency of firing that depends upon the order of spikes received and the connections' strengths. The postsynaptic potential (PSP) for a neuron i at time t is calculated as:

$$PSP(i,t) = \sum \mathrm{mod}^{\,order(j)}\, w_{j,i} \tag{1}$$

where mod ∈ (0,1) is the modulation factor, j is the index for the incoming connection and $w_{j,i}$ is the corresponding synaptic weight. When PSP reaches a given threshold (PSP_{Th}), an output spike is generated and the PSP level is reset. A detailed description of the dynamics of these neurons is given in [10].

Each individual modality has its own network of spiking neurons. In general, the output of each modality has neurons that, when issue output spikes, authenticate/not authenticate a class they represent.

Our approach for integrating modalities consists of attaching a new layer into the output of the individual modes. This layer (supramodal layer) represents the supramodal region and contains neurons that are sensitive to more than one modality [10]. In our implementation, the supramodal layer contains two spiking neurons for each class label. Each neuron representing class C in the supramodal layer has incoming excitatory connections from the output of class C neurons of each individual modality. The two neurons have the same dynamics, yet different PSP_{Th}. To one neuron, the PSP_{Th} is set in such a way that an output spike is generated after receiving incoming spike from any single modality (effectively it is a spike-based implementation of an OR gate). The other neuron has PSP_{Th} set so that incoming spikes from all individual modalities are necessary to trigger an output spike (AND gate). AND neuron maximizes the accuracy and OR neuron maximizes the recall.

In addition to the supramodal layer, a simple way to perform crossmodal coupling of modalities is designed. The crossmodal coupling is set as follows: when output neurons of an individual modality emit spikes, the spikes not only excite the neurons in the supramodal layer, but also excite/inhibit other modalities that still have ongoing processes. Effectively this excitation/inhibition influences the decision on other modalities, making it easier/more difficult to other modality to authenticate a pattern, respectively.

For the crossmodal coupling, differently from the supramodal layer connections that are only excitatory, both excitatory and inhibitory connections are implemented. Thus, class C output of one modality excites the class C neuronal maps in other modalities. On contrary, class $\hat{C}$ (not class C) output has in inhibitory effect on class C neuronal maps in other modalities.

Visual system model. The visual system is modelled with a four-layer feedforward network of spiking neurons. In [33] a single frame configuration is presented, which is extended in [34] to integrate opinions over several frames to perform authentication. Figure 2 shows the network architecture used in this work that combines opinions of being/not being a desired face over several frames. Basically, the network receives in its input several frames that are processed in a frame-by-frame manner.

The first layer (L1) neurons represent the On and Off cells of retina, enhancing the high contrast parts of a given image (high-pass filter). Second layer (L2) is

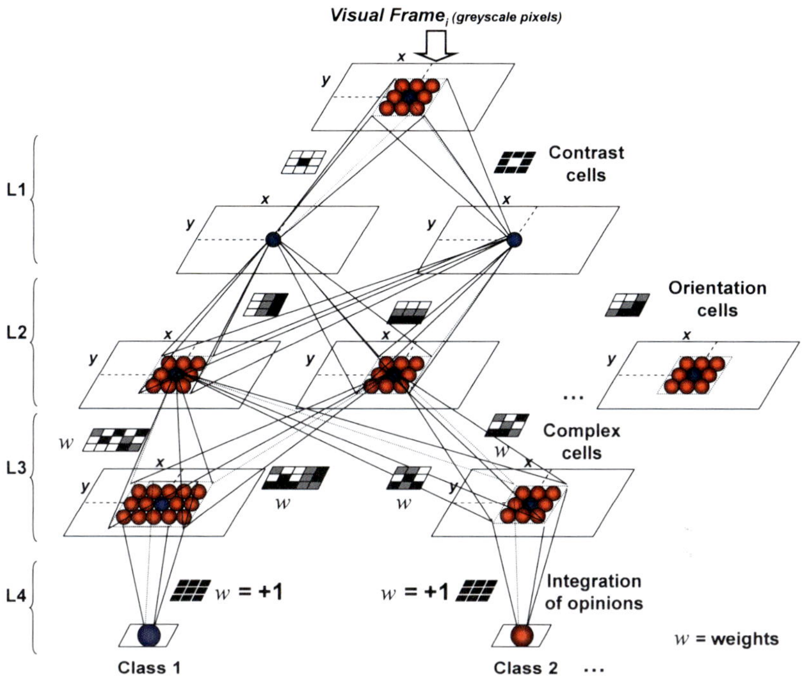

Fig. 2. SNN architecture for visual pattern recognition. Neurons in L1 and L2 are sensitive to image contrast and orientations, respectively. L3 has complex cells, trained to respond to specific patterns. It is in L3 that crossmodal coupling occurs. L4 accumulates opinions over different input excitations in time.

composed of orientation maps for each frequency scale, each one being selective to different directions. They are implemented using Gabor filters in eight directions (0°, 45°, 90°, 135°, 180°, 225°, 270°, and 315°) and two frequency scales. Maps in the third layer are trained to be sensitive to complex visual patterns (faces in our case study). In L3, neuronal maps are created or merged during learning in an adaptive on-line way. It is in L3, that neurons can receive crossmodal influences (multisensory neurons). Neurons in layer 4 (L4) accumulate opinions of being a certain class. If the opinions are able to trigger an L4 neuron to spike, the authentication is completed.

Auditory system model. The auditory system is modelled with a two layers feedforward network of spiking neurons as proposed in our previous work [35]. In short, each speaker is represented by a set of prototype vectors that compute normalized similarity scores of MFCC (Mel Frequency Cepstrum Coefficients) features considering speaker and background models. The L1 neurons that define the prototypes of a given class can be also recipients of the crossmodal excitation/inhibition (multisensory neurons). The network architecture is illustrated in Figure 3. There are two neurons in L2 for each speaker accumulating opinions over several frames of speech signals. One neuron is triggered if the speaker is authenticated and the other is triggered in case the input excitation is more likely to be a background model. This setup, despite being very simple has been proven efficient in traditional methods to tackle the short-sentence text-independent problem (typically comprised of input utterances ranging from 3 seconds to 1 minute), mainly due to the difficulty to extract and train long-term dependencies among frames [4, 22].

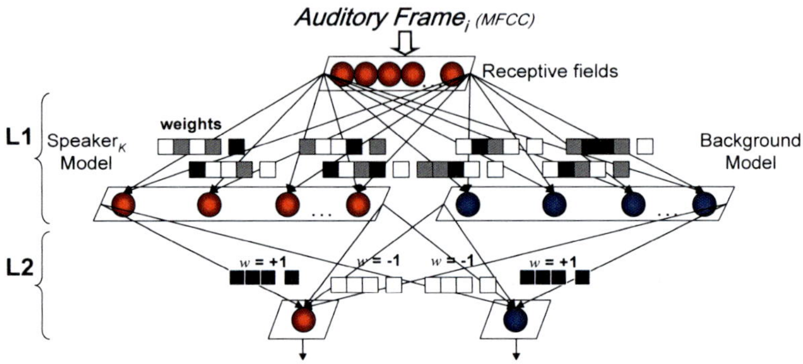

Fig. 3. Speaker authentication with spiking neural networks. L1 neurons with their respective connection weights implement the prototypes of a given class. L1 neurons also receive crossmodal excitation/inhibition. L2 accumulate binary opinions of being a claimant over several frames of speech signal.

Integration of audiovisual modalities. The detailed architecture of the audiovisual crossmodal integration is shown in Figure 4.

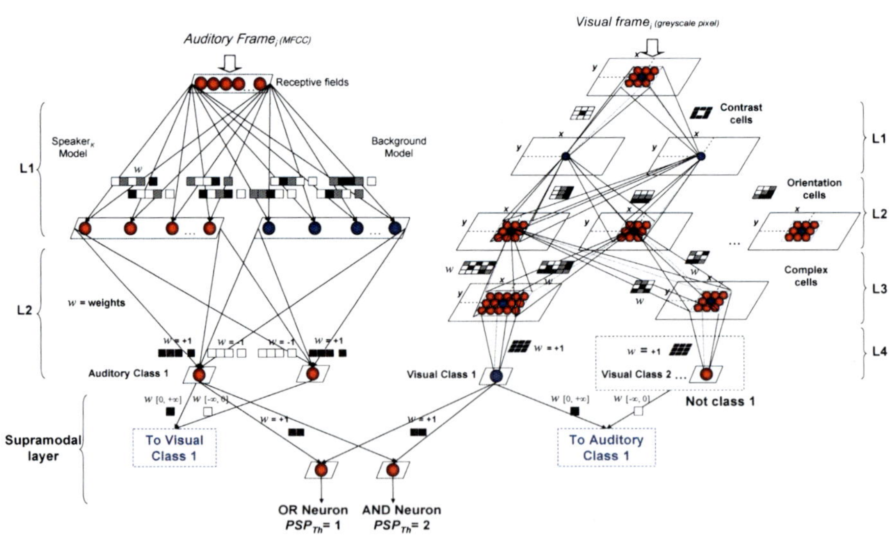

Fig. 4. Crossmodal integration of modalities using SNNs

In Figure 4 we can see the two neurons (OR and AND) in the supramodal layer. Each spiking neuron, similarly to the neurons that compose the SNNs of individual modalities, has the behaviour defined by Eq. 1. Having supramodal neurons with modulation factor mod = 1 and setting all the incoming excitatory connection weights W to 1, the PSP_{Th} that implements the OR gate for two modalities is equal to 1. The neuron implementing the AND gate receives $PSP_{Th} = 2$. Notice that, it is only possible to set deterministically these parameters because of the properties of the neurons we are using (a neuron can spike only once at any stage of computation).

In this system we effectively model the crossmodal influence through the modification of PSP_{Th} in the layers responsible for decision making within each modality. More precisely, we modify the PSP_{Th} in layer 1 (L1) neurons in the auditory model (Section 2.2) and layer 3 (L3) neurons in the visual model (Section 2.1). We use the following crossmodal parameters to denote the strength of the crossmodal influences: CM_{AVexc} (audio to video excitation), CM_{AVinh} (audio to video inhibition), CM_{VAexc} (video to audio excitation), CM_{VAinh} (video to audio inhibition), which are implemented with a proportional change in the usual PSP_{Th} values as:

$$PSP_{ThNew} = PSP_{ThOld}\,(1 + CM_{exc\,/\,inh}) \tag{2}$$

where $CM_{exc/inh}$ is negative for crossmodal excitatory influence and positive for inhibitory influence. The crossmodal influence starts from the period one individual modality produces a result and lasts until all modalities finish processing.

Notice that, in the simplest case, setting crossmodal coupling to zero, we have effectively each modality processed separately, with a simple OR/AND fusion of opinions.

3 Experimental Results

We have implemented the integration of audiovisual modalities with a network of spiking neurons and used for evaluation the VidTimit dataset [25], which contains video and audio recordings of 43 persons. Our test setup deals specifically with the audiovisual person authentication problem. Thus, a person is authenticated based on spoken phrases and the corresponding facial information as the utterance is recorded (captured in frontal view).

In the following, we present the configuration details of each individual system as well as the parameters used in integration in our experiments:

Visual: Face detection was carried out with the Viola and Jones algorithm [30] implemented in the OpenCV (http://www.intel.com/technology/computing/opencv). Faces were converted into greyscale, normalized in size (height = 60 x width = 40), convolved with an elliptical mask, and encoded into spikes using rank order coding [10]. SNN does not require illumination normalization [16]. There are two scales of On/Off cells (4 L1 neuronal maps). In scale 1, the retina filters are implemented using a 3 x 3 Gaussian grid with $\sigma = 0.9$ and scale 2 uses a

5 x 5 grid with σ = 1.5. In L2, we have 8 different directions in each frequency scale with a total of 16 neuronal maps. The direction selective filters are implemented using Gabor functions with aspect ratio $\gamma = 0.5$ and phase offset $\varphi = \pi/2$. In scale 1 we use a 5 x 5 grid and wavelength $\lambda = 5$ and $\sigma = 2.5$ and in scale 2 a 7 x 7 grid with λ and σ set to 7 and 3.5, respectively. The modulation factor for the visual neurons was set to 0.995.

Auditory: Speech signals are sampled at 16 kHz, and features are extracted using standard MFCC with 19 MEL filter sub-bands ranging from 200 Hz to 7 kHz. Each MFCC feature is then encoded into spikes using rank order coding [10]. Thus, we have 19 neurons in L1. For each speaker model, we train a specific background model. For the sake of simplicity, we use the following procedure: the background model of a speaker *i* is trained using the same amount of utterances used to train the speaker model. The utterances are randomly chosen from the remaining training speakers. We have defined *a priori* the number of neurons in the auditory L1 neuronal maps for the speaker and background model (50 neurons each). The modulation factor for auditory neurons was set to 0.9.

Integration: The crossmodal parameters were set as: $CM_{AVexc} = CM_{VAexc} = 0.1$ and $CM_{AVinh} = CM_{VAinh} = 0$. We also present the results that do not consider the crossmodal coupling, i.e., $CM_{AVexc} = CM_{VAexc} = CM_{AVinh} = CM_{VAinh} = 0$.

The system is trained to authenticate 35 users using six utterances from each user for training. To train the visual part, only two frames from each user have been used, collected while uttering two distinct phrases from the same session.

For test, we use two phrases (each phrase corresponding to one sample) recorded in two different sessions. We have 35 users x 2 samples = 70 positive claims. Simulating impostors, we use two utterances of the eight remaining users that try to break into each of the 35 users' models, which give 560 false claims.

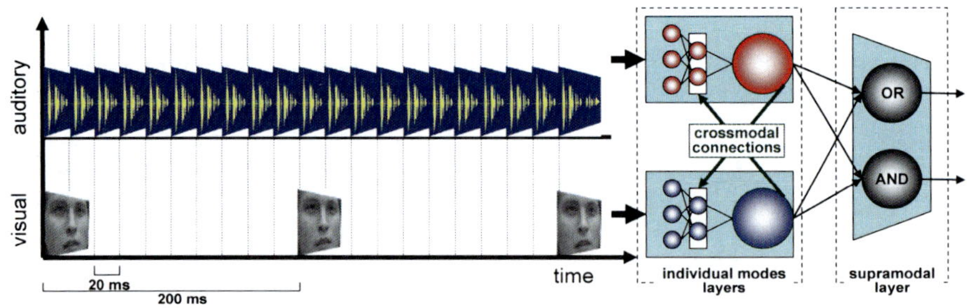

Fig. 5. Frame-based crossmodal integration of modalities using SNNs

The test is carried out frame-by-frame keeping the time correspondence between speech and visual frames. However, to speed up the computational simulations, we downsampled the visual frames (Figure 5). Five visual frames per second have been used whereas the speech samples have rate at 50 frames per second. We noticed that it does not affect the performance of the system, as for

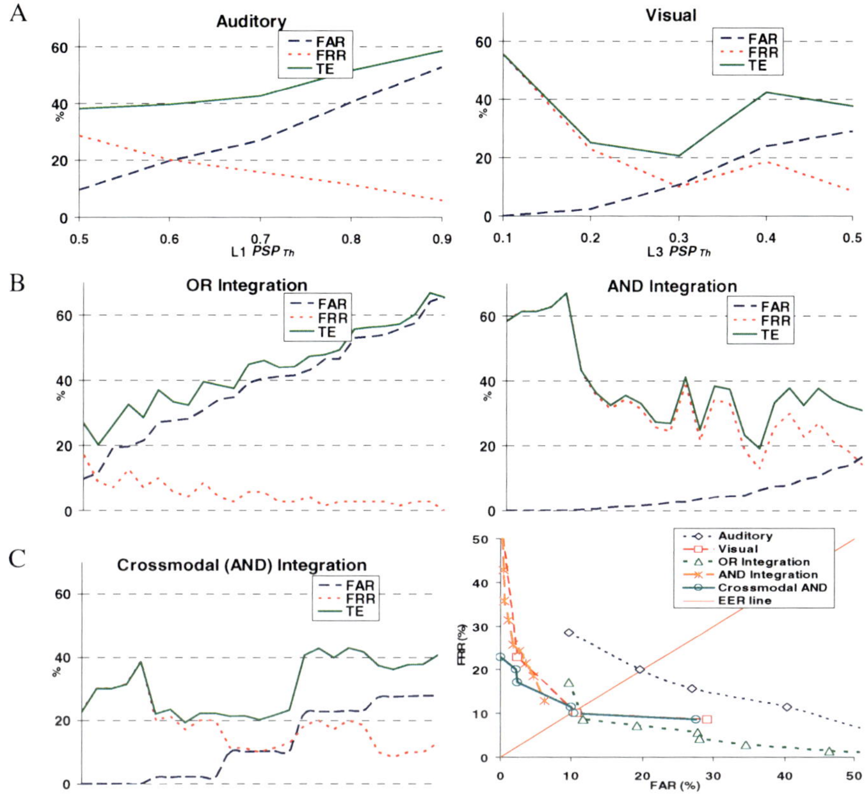

Fig. 6. A) Performance of individual modalities for different values of auditory (L1 PSP_{Th}) and visual parameters (L3 PSP_{Th}). B) Performance of the OR and AND integration of modalities with a supramodal layer of spiking neurons. (C) Performance when excitatory crossmodal influences were made active (for auditory L1 PSP_{Th} and L3 PSP_{Th} ranging from [0.5, 0.9] and [0.1, 0.5], respectively). FAR is the false acceptance rate, FRR is the false rejection rate and TE is the total error (FAR+FRR). Lowest right-hand graph: Comparison between individual modes (auditory and visual) and different integration scenarios in the FAR–FRR plane. EER is the equal error rate, where FAR = FRR.

period lower than 200 ms we could not notice substantial differences between one facial posture and another.

The supramodal layer and the crossmodal coupling are updated when an individual modality outputs a spike, which may occur once in every frame. Here, we consider the same processing time for one frame regardless of the modality, although it is well known that auditory stimulus are processed faster than a visual stimulus (difference of approximately 40 to 60 ms [28]). In our experiments, for the speech mode, the number of opinions needed to validate a person is set proportionally to the size of a given utterance (we use 20% of the total number of frames in an utterance). For the visual mode, the number of opinions needed to authenticate a person is set to two (two frames).

Figure 6A shows the best performance obtained on each individual modality. While the best total error (TE) for the face authentication is 21%, the auditory authentication is TE ≈ 38% (varying values of L1 PSP_{Th} in the auditory system and L3 PSP_{Th} in the visual system).

Figure 6B,C shows the best performance of the system considering the integration held in the supramodal layer. First, we set the crossmodal coupling parameters to zero, simulating only the OR and AND integration of individual modalities done by the supramodal layer. Then, the crossmodal coupling was made active, setting $CM_{AVexc} = CM_{VAexc} = 0.1$ and $CM_{AVinh} = CM_{VAinh} = 0$. The same parameters of individual modalities are used in this experiment, i.e., auditory parameters (L1 PSP_{Th}) and visual parameters (L3 PSP_{Th}) ranging from [0.5, 0.9] and [0.1, 0.5], respectively. The x-axis represents different combination of L1 and L3 PSP_{Th} ordered according to the performance.

Having in mind that the parameters have been optimized by hand, in Figure 7 we can see the potential advantages of the integration module. When the system needs to operate with low FAR levels (below 10%), AND and Crossmodal AND provide lower FRR than any singular modality. When the system requires operating with low FRR (below 10%), OR integration can be used instead, which gives lower FAR for the same FRR levels.

4 Conclusions and Future Directions

In this paper, we propose a new system to integrate modalities using fast spiking neurons and rank-order coding (basically time to first spike). Each individual modality is processed using specialized adaptive SNNs. Visual system extends the SpikeNet [10] with the on-line learning and multi-frame integration, however the auditory and integrated system is a novel architecture designed by us. The integration is done in a supramodal layer and one modality can influence another through a crossmodal mechanism. Under the pattern recognition perspective, we tested the network on the person authentication problem. In preliminary experiments, we can clearly see that the integration of modes enhances the performance in several operation points of the system.

In [25], the integration of modalities was explored with the VidTimit dataset using a combination of mathematical and statistical methods. The auditory system alone, using MFCC features and GMM in a noise-free setup, reached TE (total error) = FAR (false acceptance rate) + FRR (false rejection rate) ≈ 22%. The visual system was reported to have TE ≈ 8% with features extracted using PCA (principal component analysis) and SVM (support vector machine) for classification. Several adaptive and non adaptive systems to perform integration have been tested, with the best performance obtained with a new approach that builds the decision boundaries for integration considering how the distribution of opinions are likely to change under noisy conditions. The accuracy obtained with the integration reached TE ≈ 6% using 35 users for training and 8 users simulating the impostors.

Despite some differences in our experiments setup when compared to [25], our preliminary results (Fig. 6) are clearly not as good (TE never drops below 20%). There is still some room in a improving the input preprocessing and include more stages of feature extraction in our model. After all, our SNNs for auditory and visual processing are very simplified. In reality there are dozens of processing stages in the brain to deal with processing of input before information reaches the multimodal and supramodal stages. Another major issue to be explored in further improvements of our model is the coding. Rank-order code where the first spike carries the most of information has been proposed based on experiments with ultra-fast classification into two classes [29]. Here we have extended this scenario to a far more complex classification task, where in fact another coding can be more appropriate, like synchronicity or spatio-temporal pattern of spikes or perhaps a rate code [13].

The model has further constraints. It cannot take into account several biological behaviours, e.g., cannot cover familiarity decisions, semantic information, identity priming, and within and cross domain semantic priming [5, 11, 12]. In respect to the implementation, the use of frames and their respective synchronization seems to be very artificial, truncating the natural flow of information. In addition, the difference in processing time in each modality is ignored [28]. These problems need addressing through adding appropriate stages of input processing and more elaborate information coding mechanisms.

Acknowledgements

The work has been supported by the Tertiary Education Commission of New Zealand (S.G.W.) and by the NERF grant AUTX02001 funded by FRST (L.B., N.K.).

References

1. Ben-Yacoub, S., Abdeljaoued, Y., Mayoraz, E.: Fusion of face and speech data for person identity verification. Martigny-Valais-Suisse, IDIAP-RR 99-03 (1999)
2. Bimbot, F., Bonastre, J.-F., Fredouille, C., et al.: A tutorial on text-independent speaker verification. EURASIP Journal on Applied Signal Processing 7(1), 430–451 (2004)
3. Brunelli, R., Falavigna, D.: Person identification using multiple cues. IEEE Transactions on Pattern Analysis and Machine Intelligence 17(10), 955–966 (1995)
4. Burileanu, C., Moraru, D., Bojan, L., et al.: On performance improvement of a speaker verification system using vector quantization, cohorts and hybrid cohort-world models. International Journal of Speech Technology 5, 247–257 (2002)
5. Burton, A.M., Bruce, V., Johnston, R.A.: Understanding face recognition with an interactive activation model. British Journal of Psychology 81, 361–380 (1990)
6. Calvert, G.A.: Crossmodal processing in the human brain: insights from functional neuroimaging studies. Cerebral Cortex 11, 1110–1123 (2001)
7. Chevallier, S., Paugam-Moisy, H., Lemaitre, F.: Distributed processing for modelling real-time multimodal perception in a virtual robot. In: Proc. International Multi-Conference Parallel and Distributed Computing and Networks, Innsbruck, pp. 393–398 (2005)

8. Chibelushi, C.C., Deravi, F., Mason, J.S.D.: A review of speech-based bimodal recognition. IEEE Transactions on Multimedia 4(1), 23–37 (2002)
9. Crepet, A., Paugam-Moisy, H., Reynaud, E., et al.: A modular neural model for binding several modalities. In: Proc. International Conference on Artificial Intelligence (ICAI), pp. 921–928 (2000)
10. Delorme, A., Gautrais, J., van Rullen, R., et al.: SpikeNET: a simulator for modeling large networks of integrate and fire neurons. Neurocomputing 26(27), 989–996 (1999)
11. Ellis, A.W., Young, A.W., Hay, D.C.: Modelling the recognition of faces and words. In: Morris, P.E. (ed.) Modelling Cognition. Wiley, New York (1987)
12. Ellis, H.D., Jones, D.M., Mosdell, N.: Intra- and inter-modal repetition priming of familiar faces and voices. British Journal of Psychology 88, 143–156 (1997)
13. Gerstner, W., Kistler, W.M.: Spiking Neuron Models. Cambridge Univ. Press, Cambridge (2002)
14. Gonzalo, D., Shallice, T., Dolan, R.: Time-dependent changes in learning audiovisual associations: a single-trial fMRI study. NeuroImage 11, 243–255 (2000)
15. Haller, M., Hyoung-Gook, K., Sikora, T.: Audiovisual anchorperson detection for topic-oriented navigation in broadcast news. In: Proc. IEEE International Conference on Multimedia and Expo, pp. 1817–1820. IEEE, Toronto (2006)
16. Kasabov, N., Postma, E., van den Herik, J.: AVIS: a connectionist-based framework for integrated auditory and visual information processing. Information Sciences 123, 127–148 (2000)
17. Maciokas, J., Goodman, P.H.: Large-scale spike-timingdependent-plasticity model of bimodal (audio/visual) processing. Technical Report, Goodman Brain Computation Laboratory. University of Nevada, Reno (2003)
18. McIntosh, A.R., Cabeza, R.E., Lobaugh, N.J.: Analysis of neural interactions explains the activation of occipital cortex by an auditory stimulus. Journal of Neurophysiology 80, 2790–2796 (1998)
19. Messer, K., Matas, J., Kittler, J., et al.: XM2VTSDB. The extended M2VTS database. In: Proc. 2nd International Conference on Audio-Video Based Biometric Person Authentication, Washington, pp. 72–77 (1999)
20. Park, C., Choi, T., Kim, Y., et al.: Multi-modal human verification using face and speech. In: Proc. IEEE Interantional Conference on Computer Vision Systems (ICVS), pp. 54–59 (2006)
21. Poggio, T., Girosi, F.: Regularization algorithms for learning that are equivalent to multilayer networks. Science 247, 978–982 (1990)
22. Reynolds, D.A., Quatieri, T.F., Dunn, R.B.: Speaker verification using adapted Gaussian Mixture Models. Digital Signal Processing 10, 19–41 (2000)
23. Rosenberg, A.E., Soong, F.K.: Evaluation of a vector quantization talker recognition system in text independent and text dependent modes. Computer Speech and Language 2(3-4), 143–157 (1987)
24. Ross, A., Jain, A.K.: Information fusion in biometrics. Pattern Recognition Letters 24(13), 2115–2125 (2003)
25. Sanderson, C., Paliwal, K.K.: Identity verification using speech and face information. Digital Signal Processing 14, 449–480 (2004)
26. Séguier, R., Mercier, D.: Audio-visual speech recognition one pass learning with spiking neurons. In: Dorronsoro, J.R. (ed.) ICANN 2002. LNCS, vol. 2415, p. 1207. Springer, Heidelberg (2002)
27. Sharkey, A.: Combining Artificial Neural Nets: Ensemble and Modular Multi-Net Systems. Springer, New York (1999)

28. Stein, B.E., Meredith, M.A.: The Merging of the Senses. The MIT Press, Cambridge (1993)
29. Thorpe, S.J., Fabre-Thorpe, M.: Seeking categories in the brain. Science 291, 260–262 (2001)
30. Viola, P., Jones, M.: Rapid Object Detection using a Boosted Cascade of Simple Features. In: Proc. IEEE Computer Society Conference on Computer Vision and Pattern Recognition (CVPR 2001), vol. 1, pp. 511–517 (2001)
31. von Kriegstein, K., Kleinschmidt, A., Sterzer, P., et al.: Interaction of face and voice areas during speaker recognition. Journal of Cognitive Neuroscience 17(3), 367–376 (2005)
32. von Kriegstein, K., Giraud, A.: Implicit multisensory associations influence voice recognition. Plos Biology 4(10), 1809–1820 (2006)
33. Wysoski, S.G., Benuskova, L., Kasabov, N.: On-line learning with structural adaptation in a network of spiking neurons for visual pattern recognition. In: Kollias, S.D., Stafylopatis, A., Duch, W., Oja, E. (eds.) ICANN 2006. LNCS, vol. 4131, pp. 61–70. Springer, Heidelberg (2006)
34. Wysoski, S.G., Benuskova, L., Kasabov, N.: Fast and adaptive network of spiking for multi-view visual pattern recognition. Neurocomputing (2007) (under review)
35. Wysoski, S.G., Benuskova, L., Kasabov, N.: Text-independent speaker authentication with spiking neural networks. In: de Sá, J.M., Alexandre, L.A., Duch, W., Mandic, D.P. (eds.) ICANN 2007. LNCS, vol. 4669, pp. 758–767. Springer, Heidelberg (2007)

Human-Like Local Navigation System Inspired by a Hippocampal Memory Mechanism

Tsutomu Miki, Takuya Inoue, Yuta Goto, Masahiko Watanabe, and Hatsuo Hayashi

Department of Brain Science and Engineering, Graduate School of Life Science and Systems Engineering, Kyushu Institute of Technology, 2-4 Hibikino, Wakamatsu, Kitakyushu 808-0196, Japan
{miki@,inoue_takuya@edu,goto_yuta@edu}.brain.kyutech.ac.jp
{watanabe_masahiko@edu,hayashi@}.brain.kyutech.ac.jp

Abstract. We aim to develop a robust and flexible navigation system having adaptability against changes in circumstances like a human being. A practical local navigation system inspired by the structure of the entorhino-hippocampal loop circuitry is described. The local navigation system consists of a landmark extraction unit, a learning-type matching unit and a local route memory unit. In the local route memory unit, the sequence-learning mechanism of the entorhino-hippocampal system is implemented using a fully-connected-type network. This system has two operation modes. In the learning mode, a sequence of landmarks is represented by enhanced loop connections in the connection matrix of the network. In the recalling mode, the system traces the stored route comparing current landmarks with the stored landmarks using the landmark extraction and learning-type matching units. The effectiveness of the proposed system was demonstrated using an autonomous mobile robot having the proposed local navigation system.

1 Introduction

Rapid development of recent information technology brings us our comfortable daily life. One of them is navigation system which guides a route by using information of a map and a GPS. Unfortunately the system can not work at places a GPS signal does not reach and needs the latest map for accurate navigation. On the other hand, human beings can trace a route without dedicated equipments even if the circumstances changed slightly. How is a route stored in the brain? How is the stored route traced in the brain? Why does the brain cope with a slight change in circumstances? A navigation system which incorporated these brain functions enables navigation like a human being.

Igarashi et al. have proposed a network model of the entorhinal cortex layer II (ECII) with entorhino-hippocampal loop circuitry as a computational model of the sequence learning in the brain [1]. Loop connections between stellate cells in ECII are selectively potentiated by afferent signals to ECII, and consequently stellate cells

A. Hanazawa et al. (Eds.): Brain-Inspired Info. Technology, SCI 266, pp. 29–32.
springerlink.com

connected by potentiated loop connections fire successively in each theta cycle. In this paper, we propose a human-like local navigation system inspired by the sequence learning mechanism of ECII with entorhino-hippocampal loop circuitry.

2 Sequence Learning Mechanism of the Entorhinal-Hippocampal System

In the ECII network model with entorhino-hippocampal loop connections proposed by Igarashi et al. [1], a pair of afferent pulse trains to ECII whose frequencies were properly different was selected by virtue of loop connections and loop connections were selectively potentiated by those pairs of afferent signals.

Here, let's assume a route coordinated by a sequence of places A, B, C, D, and E. After signal of place A is fed into ECII, the signal is transmitted from ECII to ECV through the entorhino-hippocampal loop circuit as shown in Fig.1 (a). If a signal of place B is observed at ECII when the transmitted signal A arrived to ECV, the connection between the corresponding place cells in the ECV and the ECII is enhanced. This means that a relationship between place A and place B is established. At the next, if a signal of place C is observed at ECII when the transmitted signal B arrived to ECV, the connection between place B and place C is established in the same manner. As the results, the relationships for the following places are established in the same way as shown in Fig.1 (b).

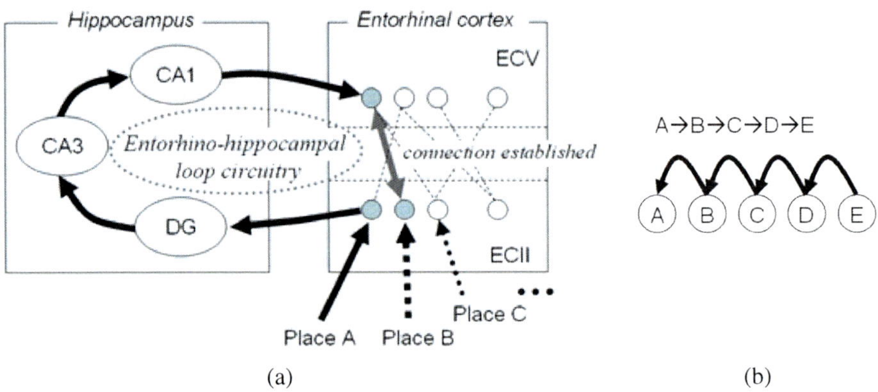

Fig. 1. Sequential learning mechanism in the entorhino-hippocampal system: (a) signal of place A is transmitted to ECV through the entorhino-hippocampal loop and is connected to signal of place B which observed after signal of place A with the loop delay, (b) a route is established by chain of connections of landmark pairs.

3 Local Navigation System Inspired by Sequential Learning Mechanism

We propose a practical navigation system inspired by the structure of the entorhino-hippocampal loop circuitry connecting ECII neurons. In this system, the sequential learning mechanism was implemented by using a fully-connected-type network, and the sequence of landmarks was embedded in the matrix of loop connections as a local route memory (Fig.2). The navigation system consists of a landmark extraction unit, a learning-type matching unit and a local route memory. Landmarks are extracted from an image obtained by a camera in the landmark extraction unit, and the learning-type matching unit specifies a current place on a route for comparing detected landmarks to the stored ones.

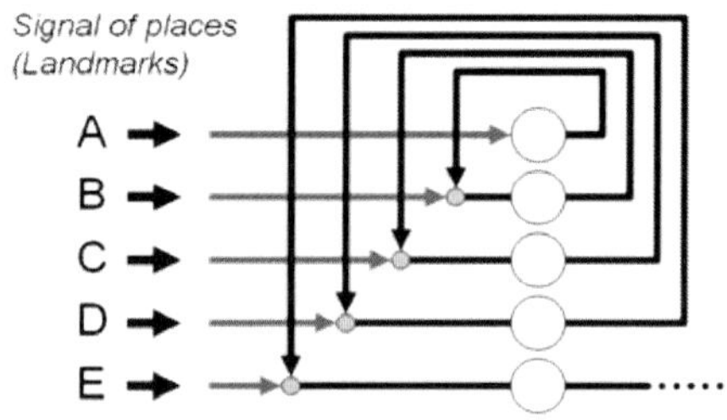

Fig. 2. Local route memory: the route is coordinated by a sequence of embedded landmarks.

The system has two operation modes, i.e. *learning mode* and *recalling mode*.

In learning mode, the system coordinates the route by a sequence of embedded landmarks through the following steps:

1. Landmark is extracted from a camera image. Here color and size of objects are used for parameters of the landmark extraction,
2. The signal of the landmark is fed into the route memory unit and the unit makes a connection between a landmark with the just before one.

By repeating the above steps, the route is memorized in the local route memory.

After that, in recalling mode, the system traces the stored route by comparing landmarks detected by a camera with the stored ones as follows:

1. Landmarks are extracted from a camera image,
2. The signals of landmarks are fed into the matching unit and compared with corresponding landmarks in the stored route. Here, a matching is done for a partial sequence of the stored local route. In this system, a partial sequence consists of four landmarks.

4 Navigation Experiments Using an Autonomous Mobile Robot

In order to confirm the effectiveness of the proposed system for real-world applications, we implement the proposed navigation system to an autonomous mobile robot "WITH" which is a basic omni-directional mobile robot platform (Fig.3 (a)) [2]. We confirmed that the system embedded in the robot extracted landmarks and stored the route in form of sequence of the landmarks. After that the robot traced the route according to the stored local route information. The proposed system which uses a partial sequence of the stored route in the route matching enables that the robot behaves like a human being to slight changes in circumstances. (Fig.3 (b)).

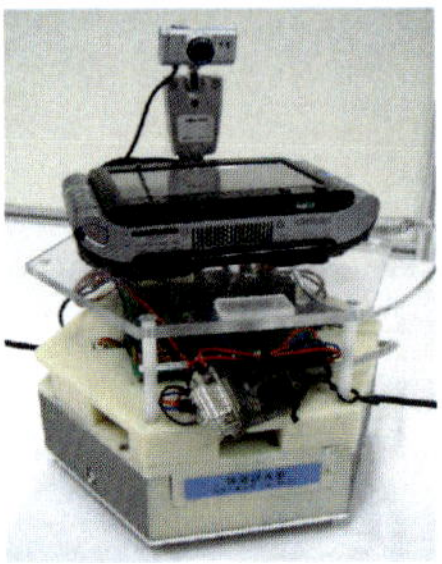

(a)

(b)

Fig. 3. (a) An autonomous mobile robot "WITH" employing the proposed navigation system: the proposed mechanism works on a palmtop PC in real-time, (b) robot traces a route autonomously according to the stored one.

5 Conclusions

Practical navigation system inspired by the sequence learning mechanism in the entorhino-hippocampal loop circuitry was described. The validity of the proposed system was confirmed by navigation experiments using an autonomous mobile robot. The proposed mechanism enables that the robot adapts like a human being flexibly to slight changes in circumstances.

Acknowledgement

This work was supported by a COE program (center #J19) granted to Kyushu Institute of Technology by MEXT of Japan.

References

1. Igarashi, J., et al.: Cogn. Neurodyn. 1, 169–184 (2007)
2. Ishii, K., Miki, T.: International Congress Series, vol. 1301, pp. 39–42 (2007)

Brain-Inspired Emergence of Behaviors Based on Values and Curiosity in Mobile Robots

Masumi Ishikawa[1], Masahiro Nagamatsu[1], Takao Hagiwara[2], Naoyuki Yamamoto[1], Fumiko Kiriake[1], Takehiko Nishida[1], Takeshi Yamakawa[1], Kazuo Ishii[1], Hideki Nakagawa[1], and Hong Zhang[1]

[1] Department of Brain Science and Engineering, Graduate School of Life Science and Systems Engineering, Kyushu Institute of Technology, 2-4 Hibikino, Wakamatsu, Kitakyushu 808-0196, Japan
ishikawa@brain.kyutech.ac.jp
[2] Quebs, Iizuka 820-0066, Japan

Abstract. We propose to introduce the desire for existence, specific curiosity, diversive curiosity, and novelty into reinforcement learning as intrinsic rewards for developing truly autonomous mobile robots capable of behaving without being told what to do. A pursuit-evasion game composed of predators and their prey is selected as a testbed. Simulation experiments and experiments using real mobile robots, WITHs, on a robotic field well demonstrate the effectiveness of introducing intrinsic rewards in addition to external rewards in the conventional reinforcement learning.

1 Introduction

Our aim is to develop truly autonomous mobile robots which are capable of behaving without being told what to do not only at a concrete level but also at an abstract level. This is what mammals do in their daily life. We believe a developmental approach is effective to realize this. The desire for existence, specific curiosity, diversive curiosity, and novelty are important factors contributing to learning and development. The present paper proposes their models.

Concerning curiosity, Shmidhuber proposed two curiosity principles in response to prediction errors and predictor improvements [1]. Oudeyer et al. proposed architecture based on the second curiosity principle [2]. Stout et al. used both external rewards and intrinsic motivation, but the latter is a kind of subgoals to accelerate learning provided by a designer [3]. In the present study, we propose to use two internal rewards of curiosity and boredom, and external rewards, and to evaluate behaviors in terms of these rewards.

2 Overview of a Task

We consider multiple predators and their prey in a robotic field. Predators learn to capture their prey, and the prey learn to escape from predators. This is done by

A. Hanazawa et al. (Eds.): Brain-Inspired Info. Technology, SCI 266, pp. 33–38.
springerlink.com

Q-learning in reinforcement learning. The desire for existence, here, is represented by keeping the level of a battery high.

When the level of its battery is higher than a threshold, an agent efficiently learns the environment motivated by curiosity[5] and novelty[6]. Usually curiosity motivates the agent to learn the environment. When a novel object is observed, however, novelty motivates the agent to approach and to closely observe it.

When the level of its battery is lower than the threshold, the predator starts to pursue its prey taking advantage of the knowledge acquired through curiosity and novelty [4]. If the speed of a predator is faster than its prey, the former learns how to pursue the latter efficiently. On the other hand, if the speed of the latter is faster than the former, cooperative behavior among predators is needed, and is expected to emerge through learning.

3 Methods for Learning

3.1 Desire for Existence

Q-learning is used for learning of both predators and their prey. Action-value function, $Q(s,a)$, represents accumulated future rewards starting from the state, s, and taking the action, a. It is updated as,

$$Q(s,a) \leftarrow Q(s,a) + \alpha[r + \gamma \max_{a'} Q(s',a') - Q(s,a)] \quad (1)$$

where α (=0.3) is a learning rate, r (=2.0) is a reward from the environment at capturing the prey, and γ (=0.99) is a discount rate. For each action, the penalty of -0.005 is given from the environment. ε-greedy method is adopted to enhance exploration.

As is well known, Q-learning has a difficulty of large computational cost. We propose, here, to decrease the number of states by various approximations. The state is defined by the battery level (20-level) and the distance to the prey (20-level). When there are obstacles, they are not enough to define the state. Therefore, the distance to an obstacle in the forward direction (10-level), and the ego-centric orientation of motion (5-level) are also added to the state definition. The speed of the predator is given by,

speed = (specified relative speed) * (maximum speed) * (motion efficiency)

where motion efficiency decreases as the battery level falls off. The prey moves slowly when the distance is large enough, but escapes at the maximum speed when the distance is less than 10.

Fig.1 illustrates the number of steps needed to capture the prey for 2 cases when there are obstacles. For simplicity, we assume one predator and one prey here. It is also assumed that only the predator learns and the prey escapes simply in the opposite direction. A simple obstacle avoidance mechanism is implemented in both the predator and the prey. Final Q-values in cases with no obstacle are used as initial Q-values to stabilize and to accelerate learning. Fig. 1 indicates that the predator captures the prey in about 35 steps using the initial Q-values, but can no longer capture the prey starting from zero Q-values.

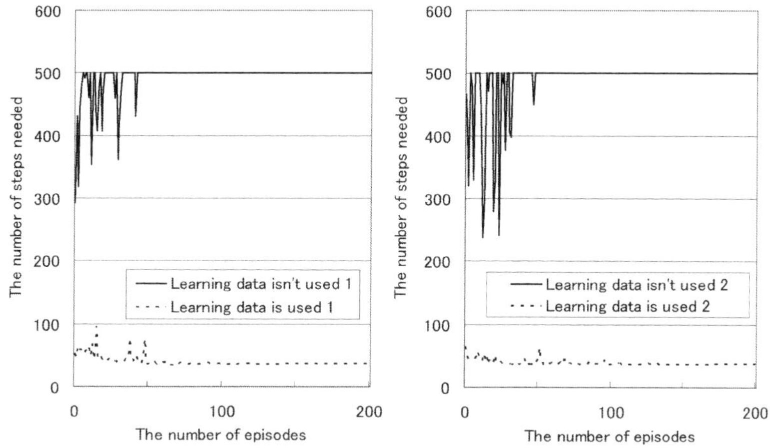

Fig. 1. The number of steps needed to capture the prey when there are obstacles.

3.2 Curiosity and Boredom

For efficient learning of the environment, we propose to introduce curiosity and boredom as internal rewards in addition to external rewards from the environment.

$$r_{\text{int}} = \delta(E_m(t-\tau) - E_m(t)) - \varepsilon(\tanh(\phi E_m(t) - 1) \tag{2}$$

$$E_m(t) = \frac{\sum_{i=0}^{\theta} e(t-i)}{\theta + 1} \tag{3}$$

$$e(t) = \| \mathbf{o}(t) - \mathbf{t}(t) \|^2 \tag{4}$$

where δ is a curiosity rate, ε is a boredom rate, τ is a time interval for evaluating learning progress, $e(t)$ is a mean prediction error(MPE) at t, E_m is a moving average of MPE over $(\theta+1)$ time steps, ϕ (=50) is a scaling parameter for $E_m(t)$, $\mathbf{o}(t)$ is the output of a predictor MLP, and $\mathbf{t}(t)$ is its teacher signal given by sensory signals.

To generate internal rewards, we develop a predictor by a multi-layer Perceptron. While the performance of the predictor improves, curiosity provides positive internal rewards. When the improvement levels off, boredom provides negative internal rewards. These rewards are used for updating the action-value function in Q-learning.

Fig. 2 illustrates the maze navigation task. At each grid, i.e., state, the agent takes one of 4 actions: up, down, right and left. The predictor learns the mapping from the current position to the current sensory signals, and generates MPE for providing internal rewards. The agent learns behaviors based on intrinsic and extrinsic rewards. When the agent reaches a goal, it receives an external reward. Otherwise the agent receives only internal rewards from curiosity and boredom.

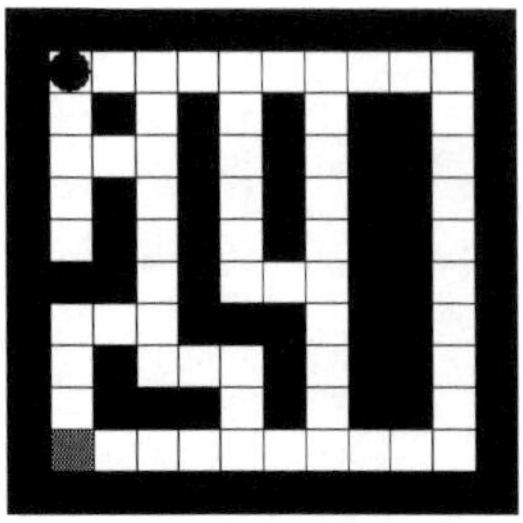

Fig. 2. The maze navigation task. The black circle: agent, the grey grid: goal.

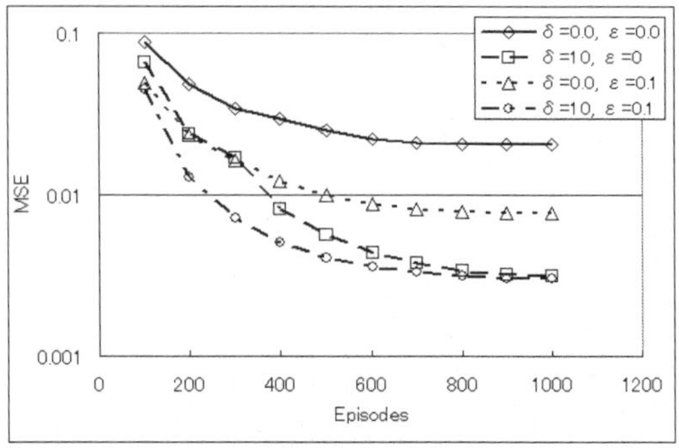

Fig. 3. Mean prediction error

We evaluate the performance of Q-learning in terms of the number of actions to reach a goal, the number of goals reached, and the MPE. Fig. 3 well demonstrates that curiosity and boredom combined gives the best performance. Providing a small negative constant reward for each action is often used in the conventional reinforcement learning to find the shortest path to a given goal. In contrast to this, we propose to use a variable negative internal reward depending on the agent's knowledge on the environment.

In case of experiments using a real mobile robot on a robotic field, the occupancy grid map is constructed based on a sequence of sensory signals. Eqs.(2)-(4) are used to calculate the improvement of a predictor using the occupancy grid map instead of an MLP.

3.3 Novelty Detection

We obtain 35-dimensional feature vectors using higher order local autocorrelation functions [7] from omni-view images taken from a mobile robot, e-puck, using a simulator, Webots. Fig. 4 illustrates the robotic field with 4 classes, i.e., straight corridor, T-junction, corner, and crossroad. Fig. 4 also depicts 21 areas. The e-puck moves randomly, taking snapshots. Fifty images are taken for each area. Accordingly, we have 600 images in the straight corridor, 200 images in the T-junction, 200 images in the corner, and 50 images in the crossroad. We try two types of data. The former is easy-to-distinguish data which does not include images near area border and images in the crossroad, and is used for training. The latter is data randomly chosen from all areas including images near area border, and is used for test.

We classify the resulting 35-dimensional feature vectors for training by the linear discriminant analysis (LDA) due to its optimality within linear computation. LDA provides a transformation matrix converting a 35-dimensional feature vector into a 2-dimensional one in case of 3 classes. We classify 2-dimensional new feature vectors for both training and test samples based on the Mahalanobis distance between a new feature vector and the centroid of each class. Feature vectors with large Mahalanobis distance are regarded as novel.

3	1	2	1	3
1		1		1
2	1	+	1	2
1		1		1
3	1	2	1	3

Fig. 4. The robotic field with 4 classes: straight corridor(1), T-junction(2), corner (3), crossroad(+).

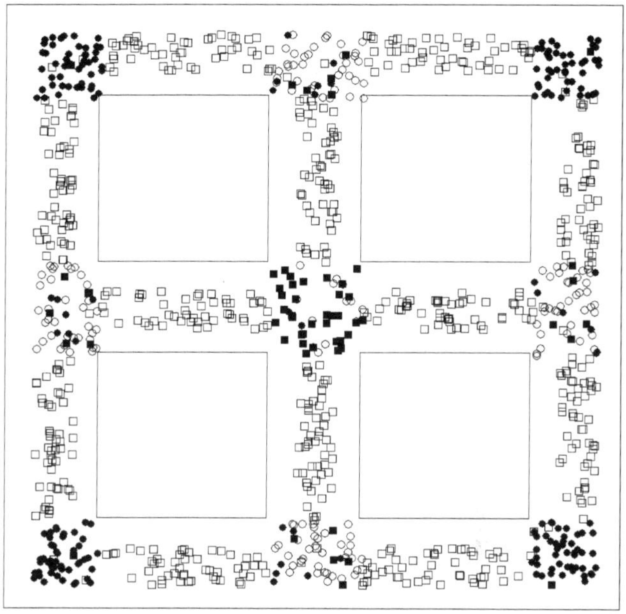

Fig. 5. The resulting classification for test data. The square, circle, filled circle and filled square stand for straight corridor, T-junction, corner and outliers, respectively.

3.4 Emergence of Cooperative Behaviors

As mentioned in Section 2, if the speed of the prey is faster than the predator, cooperative behavior among predators is needed. Assuming this and the existence of obstacles, we consider here a pursuit game in which two predators pursue one prey. In this case, the conventional reinforcement learning faces difficulty in capturing the prey, because a combination of pursuing and ambushing is required. In such a case, necessary number of actions to capture the prey becomes large, and sophisticated cooperation would be required. Therefore the conventional reinforcement learning cannot generate a sequence of reasonable actions, especially at an early stage of learning. We propose a reinforcement learning algorithm which starts with a primitive hand-crafted algorithm to ensure external rewards for rapid learning, and continuously improves it at later stages.

4 Conclusions and Discussions

We performed simulation experiments on reinforcement learning of a predator and its prey, and on the effectiveness of introducing curiosity and boring as internal rewards. We also carried out preliminary experiments on novelty detection and emergence of cooperative behaviors. Skill acquired through learning was implemented on real mobile robots, WITHs. Simulation experiments and experiments using real mobile robots on a robotic field well demonstrated the effectiveness of introducing the desire for existence, specific curiosity, diversive curiosity, and novelty as intrinsic rewards in addition to external rewards in the conventional reinforcement learning.

What was done, however, is only a first step towards truly autonomous mobile robots. Whether or not the proposed method really outperforms the conventional methods has not seriously been evaluated. Whether or not selection of behaviors based on values and curiosity really leads to better performance in the long term has not been evaluated either. Much remains for future study.

Acknowledgements

This research was supported by a COE program (#J19) granted to Kyushu Institute of Technology by the Ministry of Education, Culture, Sports, Science and Technology(MEXT), Japan. It was also supported by Grant-in-Aid for Scientific Research(C)(18500175) from MEXT, Japan.

References

1. Shmidhuber, J.: Self-motivated development through rewards for predictor errors/ improvements. In: Developmental Robotics, 2005 AAAI Spring Symposium (2005)
2. Oudeyer, P.-Y., Kaplan, F., Hafner, V.V.: Intrinsic Motivation Systems for Autonomous Mental Development. IEEE Trans. EC 11(2), 265–286 (2007)
3. Stout, A., Konidaris, G.D., Barto, A.G.: Intrinsically motivated reinforcement learning: A promising framework for developmental robot learning. In: Developmental Robotics AAAI Spring Symp. (2005)
4. Hagiwara, T., Ishikawa, M.: Emergence of Behaviors Based on the Desire for Existence. Brain IT 2007, 41 (2007)
5. Yamamoto, N., Ishikawa, M.: Curiosity and boredom based on prediction error as novel internal rewards. Brain IT 2007, 43 (2007)
6. Kiriake, F., Ishikawa, M.: Classfication and novelty detection of omni-view images taken from a mobile robot. Brain IT 2007, 42 (2007)
7. Otsu, N., Kurita, T.: A new scheme for flexible and intelligent vision systems. In: IAPR Workshop on Computer Vision, Tokyo, pp. 431–435 (1988)

Emergence of Behaviors by Reinforcement Learning Based on the Desire for Existence

Takao Hagiwara[1] and Masumi Ishikawa[2]

[1] Cueβs Co., Ltd., Research and Development Division,
303 FSC 526-1 Koubukuro, Iizuka City, Fukuoka 820-0066, Japan
hagiwara@cuebs.co.jp
[2] Department of Brain Science and Engineering, Graduate School of Life Science and Systems Engineering, Kyushu Institute of Technology,
2-4 Hibikino, Wakamatsu, Kitakyushu 808-0196, Japan
ishikawa@brain.kyutech.ac.jp

Abstract. We propose to introduce the desire for existence into reinforcement learning as intrinsic motivation for developing robots with emerging behaviors. A pursuit-evasion game composed of predators and their prey is selected as a testbed. To demonstrate the effectiveness of using the desire for existence, simulation experiments and experiments using mobile robots, WITH, on the robotic field are carried out.

1 Introduction

Our aim is to develop truly autonomous mobile robots which are capable of behaving without being told what to do not only at a detailed level but also at an abstract level. This is what mammals do in their daily life. We believe a developmental approach is effective to realize this. The desire for existence, specific curiosity, diversive curiosity, novelty and so forth are important factors contributing to learning and development. The present paper focuses on how the desire for existence enables emergence of behaviors.

2 Overview of a Task

We consider predators and its prey on a robotic field. Predators learn to capture their prey, and the prey learn to escape from predators. This is done by Q-learning in reinforcement learning. The desire for existence, here, is represented by keeping the level of a battery higher than a given threshold. A typical example of a predator is a carnivore and that of a prey is a herbivore. Hereafter, these terms are used interchangeably.

When the level of its battery is higher than a threshold, an agent learns the environment motivated by curiosity [1] and novelty [2]. When the level of its battery is lower than the threshold, it starts to pursue its prey taking advantage of the knowledge it acquired with curiosity and novelty. For simplicity, we assume one predator and one prey in this paper. Skill acquired through learning is implemented on real mobile robots to demonstrate its effectiveness on the robotic field.

A. Hanazawa et al. (Eds.): Brain-Inspired Info. Technology, SCI 266, pp. 39–44.
springerlink.com

3 Methods for Learning

We adopt Q-learning for the learning of both the predator and the prey. Action-value function, $Q(s,a)$, represents accumulated future rewards starting from the state, s, and taking the action, a. Action-value function is updated as,

$$Q(s,a) \leftarrow Q(s,a) + \alpha[r + \gamma \max_{a'} Q(s',a') - Q(s,a)] \quad (1)$$

where α (=0.3) is a learning rate, r(=2.0) is a reward from the environment at successful capturing of the prey, and γ(=0.99) is a discount rate. For each action penalty of -0.005 is given. ε-greedy is adopted.

As is well known, Q-learning has a difficulty of large computational cost. Various methods have been proposed to overcome this difficulty. In this paper, we propose to decrease the number of states as follows.

State is defined by the battery level (20-level) and the distance to the prey (20-level) in the simulation experiments. When there are obstacles, the distance and the battery level are not enough to define the state. In addition to these, the distance to an obstacle in the forward direction (10-level), and the ego-centric orientation of movement (5-level) are added to the state definition.

Each agent has parameters: battery level, movement efficiency (decrease in battery level for unit movement), and the maximum speed. The speed of an agent is defined by,

speed = specified speed * maximum speed * remaining battery level

The herbivore, on the other hand, does not learn and simply escape when the carnivore approaches. When the distance is large enough, it moves slowly, and when the distance becomes less than 10, it starts to escape at the maximum speed.

When the level of a battery becomes lower than a threshold, the predator starts pursuing. Generally speaking, if the predator captures the prey when its battery level is significantly higher than the threshold, it is not efficient enough. Therefore, it is advantageous to learn the threshold by reinforcement learning. This can be done by providing punishment when the predator captures the prey too early.

4 Simulation Experiments

4.1 Pursuit without Obstacles

To realize infinite field in simulation, the upper and the lower edges are connected, and the right and the left edges are connected. The following three cases are tried.

Case 1. Only the predator learns, and the prey escapes simply in the opposite direction.

Fig. 1 illustrates the number of steps needed for the predator to capture its prey. As learning proceeds, the predator captures its prey in about 150 steps.

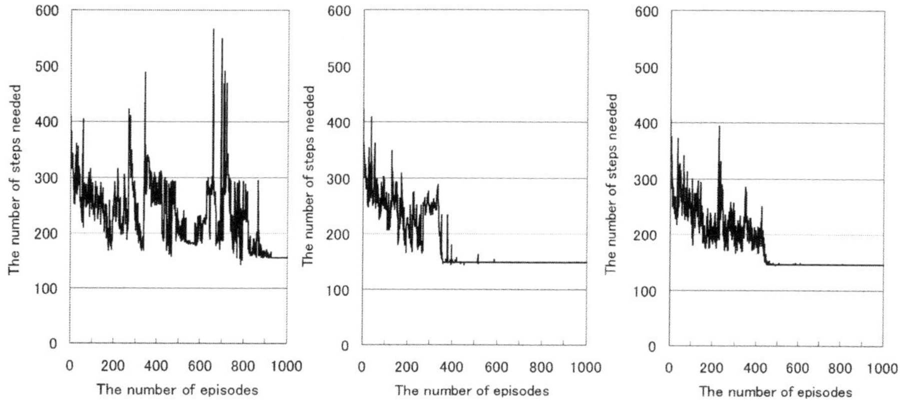

Fig. 1. The number of steps needed. Three different stands for three trials

Case 2. Only the predator learns and the prey escapes in the random direction.

Fig. 2 illustrates the number of steps needed for the predator to capture its prey. As learning proceeds, the predator captures its prey in about 90 steps, which is smaller than that in Case 1. This is due to inefficient escape of its prey from the predator.

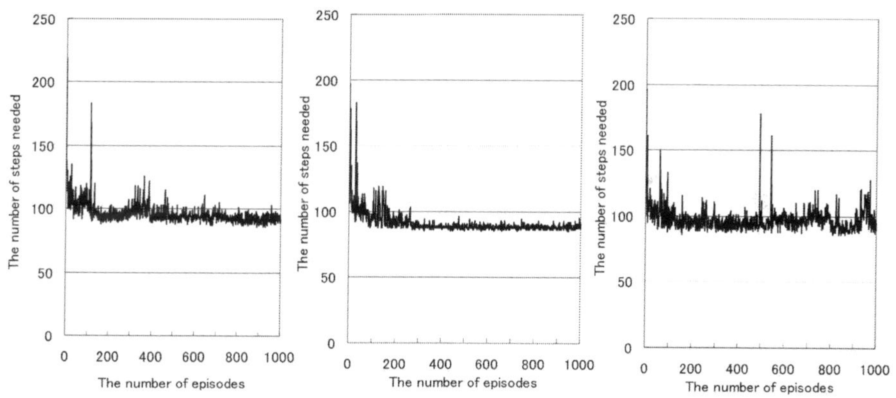

Fig. 2. The number of steps needed in Case 2

Case 3. Both the predator and the prey learn simultaneously.

Fig. 3 illustrates the number of steps needed for the predator to capture its prey. The number of steps needed is not stable due to imperfect learning. Even during the last 500 episodes, the predator captures its prey in about 3/4 of episodes. Its improvement is left for our immediate future study.

4.2 Pursuit with obstacles

Only the predator learns and the prey escapes in the opposite direction in the robotic field with obstacles. A simple obstacle avoidance mechanism is imple-mented into

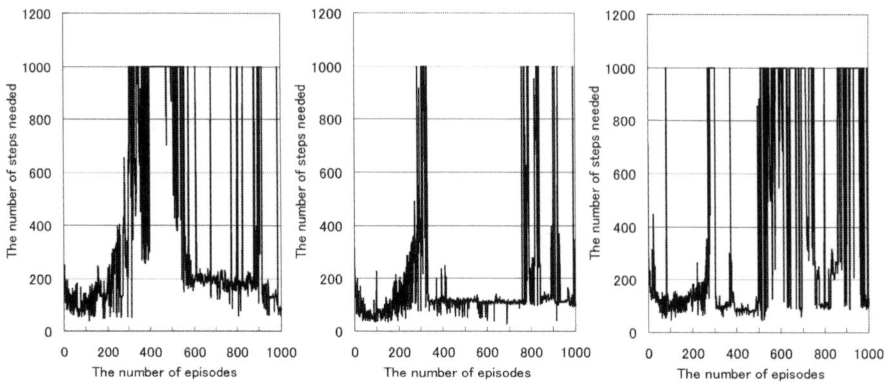

Fig. 3. The number of steps needed in Case 3

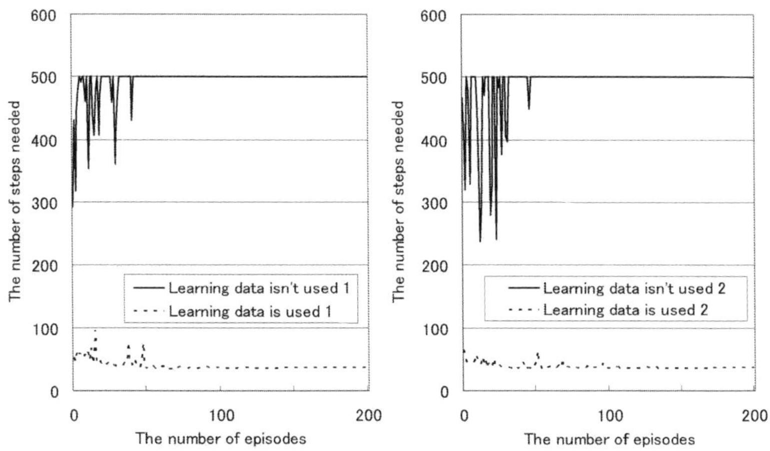

Fig. 4. The number of steps needed to capture the prey

the predator and the prey. Since the number of states and actions increases in the presence of obstacles, computational cost increases. Accordingly, final Q-values in Case 1 are used as initial Q-values to stabilize and accelerate learning. Fig. 4 illustrates that the predator captures the prey in about 35 steps with this initial Q-values, but can no longer capture the prey without them.

4.3 Learning of the Threshold for Pursuit

When the predator captures the prey with high battery level, a negative reward is provided to suppress early capture. On the other hand, when the predator captures the prey with low battery level, a positive reward is provided. Owing to these rewards, the predator is able to learn the appropriate threshold for pursuit. Fig. 5 illustrates that the remaining battery level is about 35% compared with about 75% in Sect 4.2. Fig. 6 illustrates Q-values for these two cases.

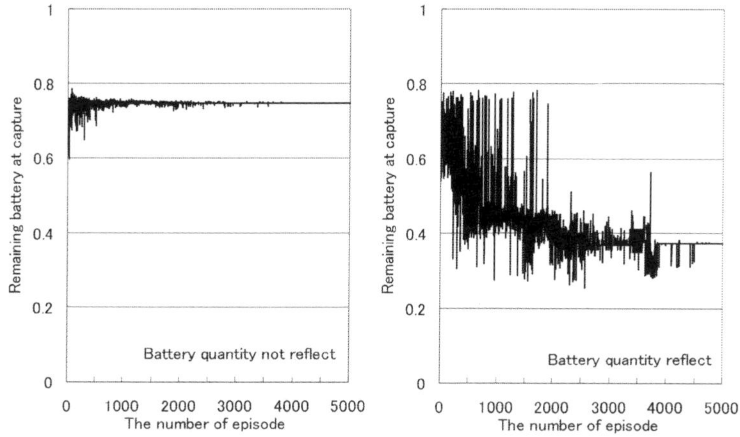

Fig. 5. The level of remaining battery at the time of capture

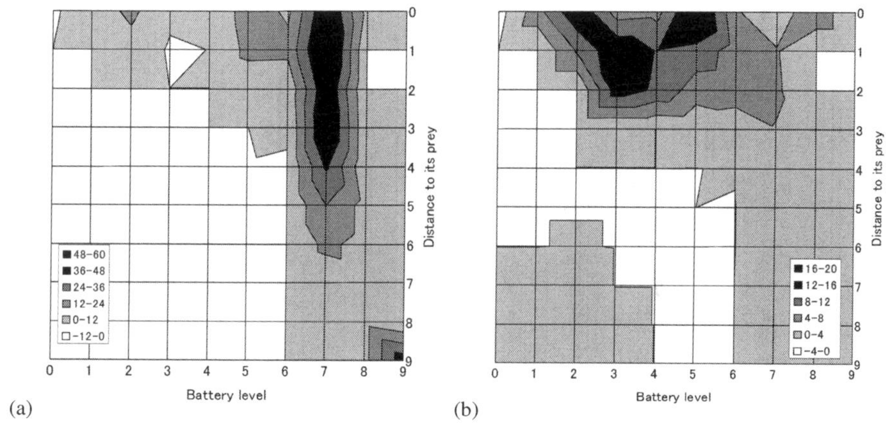

Fig. 6. Q-values, (a) Battery level not trained, (b) Battery level trained.

5 Conclusions and Discussions

We proposed to introduce the desire for existence into reinforcement learning for developing robots with emerging behaviors. To demonstrate the effectiveness of using the desire for existence, simulation experiments on a pursuit-evasion game composed of predators and their prey was studied and demonstrated its effectiveness.

Simulation results were implemented on real mobile robots, WITH. In real robots, the distance to the other robot and its ego-centric orientation are required, and are provided by omni-vision camera on the robot.

Acknowledgement

This research was partially supported by a COE program (center #J19) granted to Kyushu Institute of Technology by MEXT, Japan.

References

1. Yamamoto, N., Ishikawa, M.: Curiosity and boredom based on prediction error as novel internal rewards. Brain Inspired IT 2007, 43 (2007)
2. Kiriake, F., Ishikawa, M.: Classfication and novelty detection of omni-view images taken from a mobile robot. Brain Inspired IT 2007, 42 (2007)

Classification and Novelty Detection of Omni-view Images Taken from a Mobile Robot

Fumiko Kiriake[1] and Masumi Ishikawa[2]

[1] Department of Brain Science and Engineering, Graduate School of Life Science and Systems Engineering, Kyushu Institute of Technology,
2-4 Hibikino, Wakamatsu, Kitakyushu 808-0196, Japan
kiriake-fumiko@edu.brain.kyutech.ac.jp
[2] Department of Brain Science and Engineering, Graduate School of Life Science and Systems Engineering, Kyushu Institute of Technology,
2-4 Hibikino, Wakamatsu, Kitakyushu 808-0196, Japan
ishikawa@brain.kyutech.ac.jp

Abstract. We propose to use novelty as one of intrinsic motivations for learning in developmental robotics. Our approach classifies omni-view images taken from a mobile robot, finds outliers, and detects novelty. We use linear discriminant analysis for classification due to its optimality within linear computation. Experimental results demonstrate that although there are many misclassifications, there's a possibility of making a new class composed of novel data.

1 Introduction

Novelty Filter, Kohonen and Oja proposed in 1976[3], is famous method to detect novelty. One feature of Novelty Filter is good for notice new data, and another feature is to be used only to detect novelty. To control robots, we need information not only whether there is novel data but also which conditions the robot is. Hang and Weng proposed Hierarchical discriminant Regression (HDR) in 2000[1], which allow classify data and detect novelty using the tree architecture. They tried easy problem to detect novel data, but they experimented classification problem using images took in corridor. It might be able to use for novelty detection with corridor images. We classify features of images and use simple methods.

2 Feature Extraction from Images

We propose to use higher-order local autocorrelation functions (HLAC) [4] to extract features from images, which is an extension of the autocorrelation function. HLAC is defined as,

$$\begin{aligned} &\mathbf{x}(\mathbf{a}_1,\mathbf{a}_2,\ldots,\mathbf{a}_N) \\ &= \int f(\mathbf{r})f(\mathbf{r}+\mathbf{a}_1)\ldots f(\mathbf{r}+\mathbf{a}_N)d\mathbf{r} \end{aligned} \tag{1}$$

where f($\mathbf{r}$) is the brightness at pixel, r, and $\mathbf{a}_\mathrm{i}$ is i-th displacement vector. We adopt 2 as the order of correlation, and 3x3 as a domain for integration. Since

A. Hanazawa et al. (Eds.): Brain-Inspired Info. Technology, SCI 266, pp. 45–49.
springerlink.com

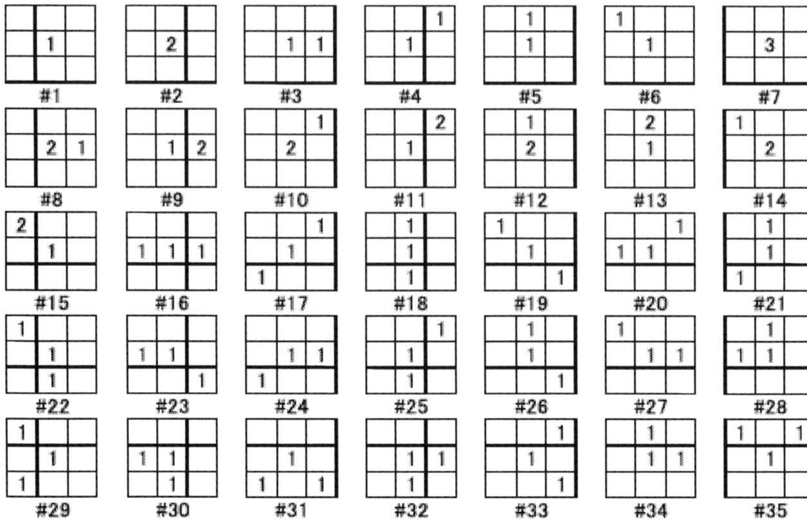

Fig. 1. The mask patterns for gray scale images, adapted from [2]

there are 35 pairs of displacement vectors, the corresponding 35 mask patterns in Fig. 1 generate a 35-dimensional feature vector from a gray scale omni-view image taken from a mobile robot. The resulting feature vectors are used for samples in classification.

3 Simulation Experiments

A software simulator, Webots, is used to acquire images from the mobile robot, e-puck. Fig. 2 illustrates the robotic field on Webots with 4 classes, i.e., straight corridor, T-junction, corner, and crossroad. Fig. 2 also depicts 21 areas. The e-puck moves randomly, taking snapshots. Fifty images are taken for each area. Accordingly, we have 600 images in the straight corridor, 200 images in the T-junction, 200 images in the corner, and 50 images in the crossroad. We try two types of data. The former is easy-to-distinguish data which does not include images near area border and images in the crossroad, and is used for training. The latter is data randomly chosen from all areas including images near area border, and is used for test.

We apply linear discriminant analysis (LDA) to the resulting 35-dimensional feature vectors for training due to its optimality within linear computation. There are any data which is useless for classification in 35-dimensional feature vectors. We can pay attention for useful data using LDA. LDA provides a transformation matrix composed of eigenvectors, which converts a 35-dimensional feature vector into a 2-dimensional one in case of 3 classes. We classify 2-dimensional feature vectors based on the Mahalanobis distance between each feature vector and the centroid of each class. Feature vectors with large Mahalanobis distance are regarded as novel.

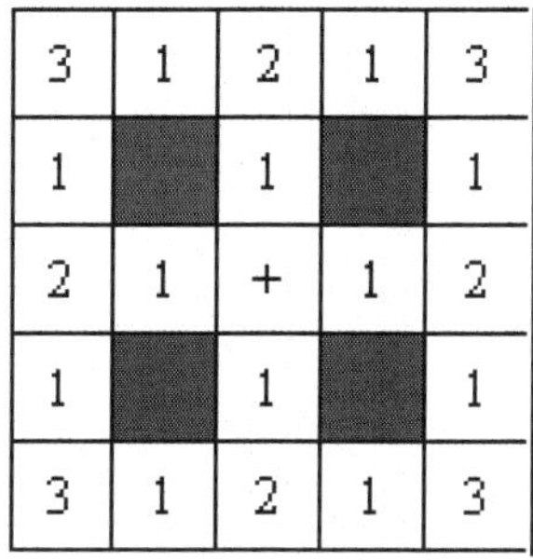

Fig. 2. The robotic field with 4 classes: straight corridor (1), T-junction (2), corner (3), crossroad (+)

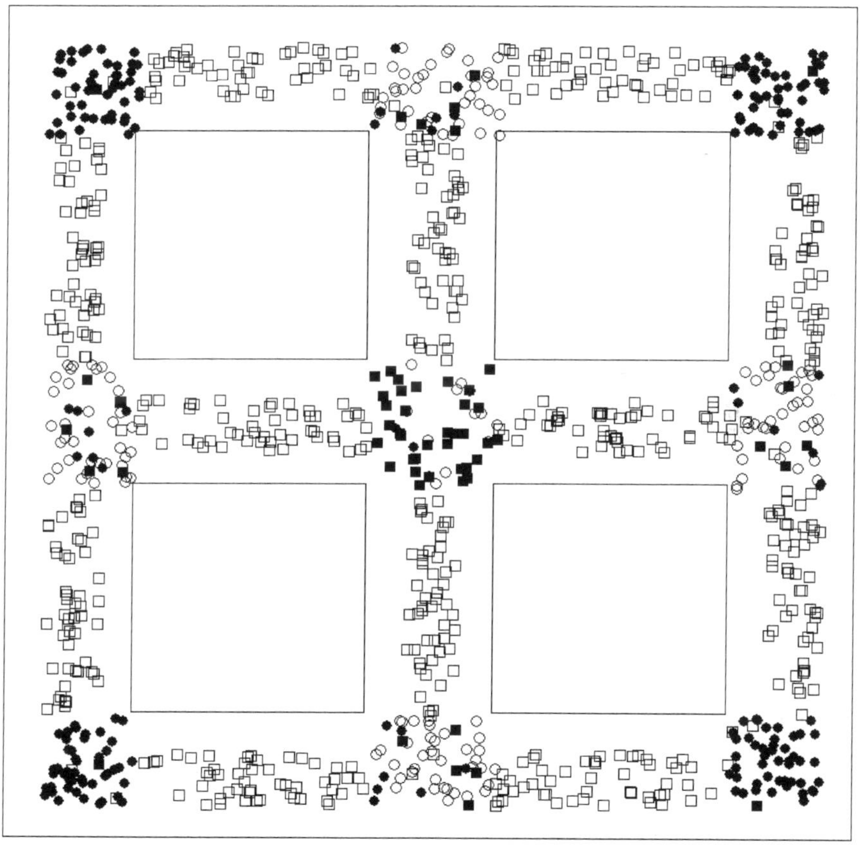

Fig. 3. The resulting classification for test data. The square, circle, filled circle and filled square stand for straight corridor, T-junction, corner and outliers, respectively.

4 Results

Fig. 3 depicts the resulting classification of images in the robotic field. The data in the crossroad assumed to be correct when classified as outliers. Supposing outliers in Fig. 3 form a novel class, most of the data classified as novel are in the cross road area. If outliers can be treated as "novel," we make a new class in the crossroad area.

The training data generate no misclassification, and the test data generate 83 misclassifications, which is 7.9% among 1050 images. This result is not satisfactory, and needs to be improved.

5 Conclusions

We classified omni-view images taken from a mobile robot by the linear discriminant analysis based on 35-dimentional feature vectors. There are 83 misclassifications (7.9%) in test data. They are attributed to the so-called perceptual aliasing, inherent difficulty due to samples near area border, and so forth. The first one suggests the need for a classification based on a sequence of images. The appropriate setting of the threshold which generates smaller number of misclassifications is also left for immediate future study. Other types of sensory data such as distance and color information can be combined with PHLAC. We did a preliminary study on novelty detection based on color information in real images.

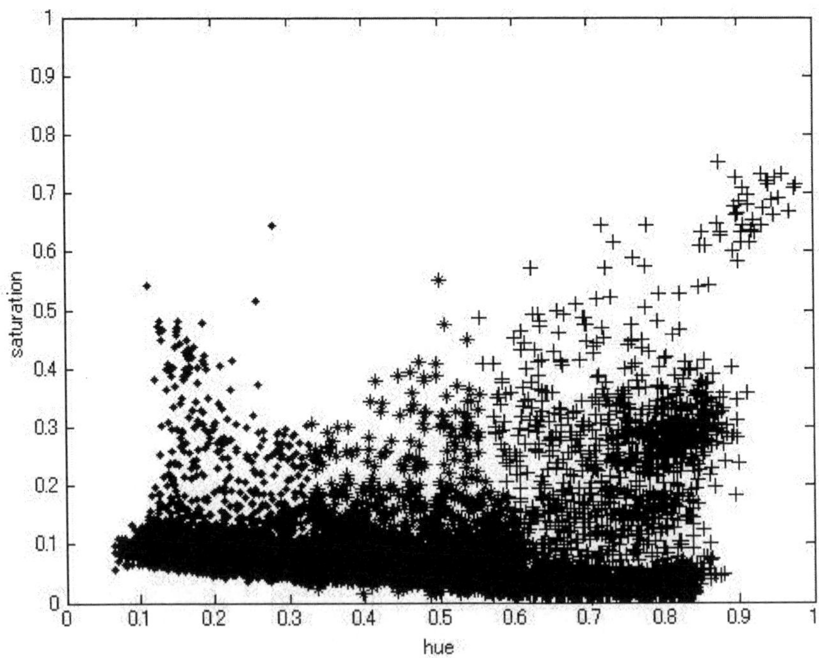

Fig. 4. The training hue and saturation value classified by K-means

We use hue and saturation values for novelty detection. Training data contained colors of environment and robot. The training color value was classified by K-means. The robot was moving in the environment and took images to detect novel. If we let object with new color, the robot notice the novelty and make new class. Fig.4 depicts the hue and saturation value. The left side data was yellow, the right side data was red, and the data at the bottom is white. It might be difficult to classify white color data because white can be seen any color with a little difference of right. We might have to think of the distribution for making class from training data.

Acknowledgement

This research was supported by a COE program (center #J19) granted to Kyushu Institute of Technology by MEXT, Japan.

References

1. Weng, J., Hwang, W.-S.: Incremental Hierarchical Discriminant Regression. IEEE Transactions on Neural Networks 18(2) (March 2007)
2. Hotta, K., Kurita, T., Mishima, T.: Scale invariant face detection method using higher-order local autocorrelation features extracted from log polar image. In: Third International Conference of Face and Gesture Recognition, pp. 70–75 (1988)
3. Kohonen, Oja: First adaptive formation of orthogonalizing filters and associative memory in recurrent networks of neuron-like elements. Biological Cybernetics, 85–95 (1976)
4. Otsu, N., Kurita, T.: A new scheme for flexible and intelligent vision systems. In: IAPR Workshop on Computer Vision, Tokyo, pp. 431–435 (1988)

Curiosity and Boredom Based on Prediction Error as Novel Internal Rewards

Naoyuki Yamamoto* and Masumi Ishikawa

Department of Brain Science and Engineering, Graduate School of Life Science and Systems Engineering, Kyushu Institute of Technology,
2-4 Hibikino, Wakamatsu, Kitakyushu 808-0196, Japan
yamamoto-naoyuki@edu.brain.kyutech.ac.jp,
ishikawa@brain.kyutech.ac.jp

Abstract. In this paper, the use of two internal reward models, curiosity and boredom, is proposed. Experiments on a maze navigation task demonstrated that appropriate values of parameters simultaneously improved the performance of the predictor of the environment and increase the external rewards compared with the conventional reinforcement learning. In conclusions, the relation between the proposed method and active learning, diversive curiosity, and specific curiosity is also discussed.

1 Introduction

Recently, entertainment robots are readily available for use at home. Since their behaviors are mainly preprogrammed by their designers, they are not able to respond appropriately in unprepared situation. To overcome this difficulty, we believe a developmental approach is promising and contributes to emergence of intriguing behaviors of mobile robots. Developmental robotics is an emerging field toward truly autonomous robots.

Reinforcement learning is suitable for learning of behaviors of a mobile robot based on the interaction between an agent and an environment [1]. Shmidhuber proposed two curiosity principles inspired by the behavior of mammals that generates curiosity reward in response to prediction errors and predictor improvements [2]. Oudeyer and Kaplan et al. proposed architecture based on the second curiosity principle implemented in machine predictor, i.e. multi-layer Perceptron[3].

Although Shmidhuber mentioned a reward system with external and internal rewards, only internal reward was actually used in the experiments [2]. The learning model in [3] is reinforcement learning with only intrinsic motivation. Barto and his colleagues used both external rewards and intrinsic motivation, but the latter is a kind of subgoals to accelerate learning provided by a designer [4].

In the real world, however, an agent decides its behavior based on both intrinsic and extrinsic motivation. In this study, we propose to simultaneously use two

* Corresponding author.

A. Hanazawa et al. (Eds.): Brain-Inspired Info. Technology, SCI 266, pp. 51–55.
springerlink.com

internal rewards of curiosity and boredom, and external rewards, and to evaluate behaviors in terms of these rewards.

2 Method

Fig. 1 illustrates the architecture for both internal and external rewards. An external reward is given when reaching a goal, and an internal reward is provided by a curiosity model (right-hand in Fig1). They are used for updating an action-value function in Q-learning.

$$r = r_{ext} + r_{int} \tag{1}$$

where r_{ext} is an external reward provided by goal reaching, and r_{int} is an internal reward. To generate internal rewards, we develop a predictor by a multi-layer Perceptron. The resulting mean square error (MSE) provides internal rewards. While the performance of the predictor is improving, curiosity provides positive internal rewards, and contributes to further improvement of the predictor, i.e., the decrease in MSE. When the improvement levels off, boredom provides negative internal rewards. These rewards are used for updating an action-value function in Q-learning.

$$r_{int} = r_{curiosity} + r_{boredom} \tag{2}$$

$$r_{curiosity} = \begin{cases} \delta(E_m(t-\tau) - E_m(t)) & if \quad E_m(t-\tau) > E_m(t) \\ 0 & if \quad E_m(t-\tau) \le E_m(t) \end{cases} \tag{3}$$

$$r_{boredom} = \varepsilon(tanh(\phi \cdot E_m) - 1.0) \tag{4}$$

where δ and ε are coefficients of positive(curiosity) and negative(boredom) internal rewards, respectively, and ϕ is a parameter controlling the amount of

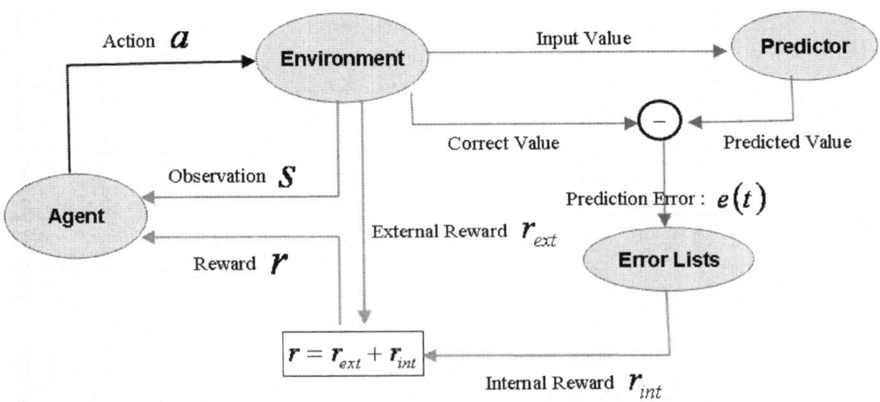

Fig. 1. The proposed architecture, introducing external and internal rewards in updating Q-function

boredom. In this experiment ϕ takes the value of 50. τ is a time window parameter, and $E_m(t)$ is a moving average of MSE defined by

$$E_m(t) = \frac{\sum_{i=0}^{\theta} e(t-i)}{\theta + 1} \tag{5}$$

where θ is a smoothing parameter in [3] and takes the value of 5 here.

$$e(t-i) = \left\| T(t-i) - O(t-i) \right\|^2 \tag{6}$$

where $O(t-i)$ is the output of MLP and $T(t-i)$ is the corresponding teacher signal.

3 Experiments

Fig. 2 illustrates the environment for the maze navigation task. The state of the agent is the current grid. The agent takes one of 4 actions, i.e., up, down, right and left. The predictor learns the mapping from the current position to the current sensory information, and generates MSE for providing internal rewards. In this experiment, the agent learns behaviors based on intrinsic and extrinsic rewards. When the agent reaches a goal, it receives a fixed amount of external reward, +1. Otherwise the agent receives only internal rewards from curiosity and boredom models.

We first evaluate the performance of Q-learning in terms of the number of actions to reach a goal (Fig. 3), the number of goals reached (Fig. 4), and the MSE (Fig. 5). Figs. 3 and 4 depict that the case with only negative internal rewards, i.e., boredom, demonstrates the best performance in terms of the number of actions to reach a goal and the number of goals reached. Providing a small negative constant reward for each action of the agent is often used in the conventional RL to find the shortest path to a given goal. In contrast to this, we propose to use a negative internal reward depending on the agent's knowledge of the environment. In other words, the more

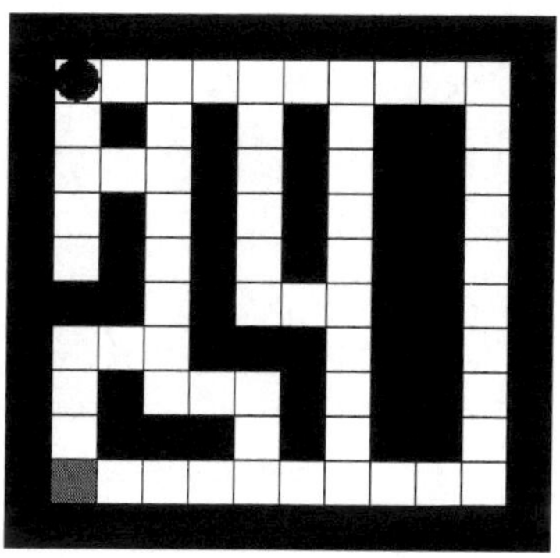

Fig. 2. The maze navigation task. The black circle and the grey grid represent the agent and the goal, respectively.

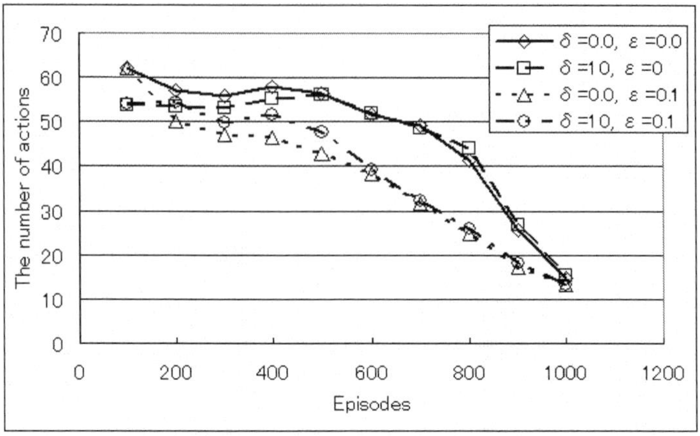

Fig. 3. The number of actions to reach a goal

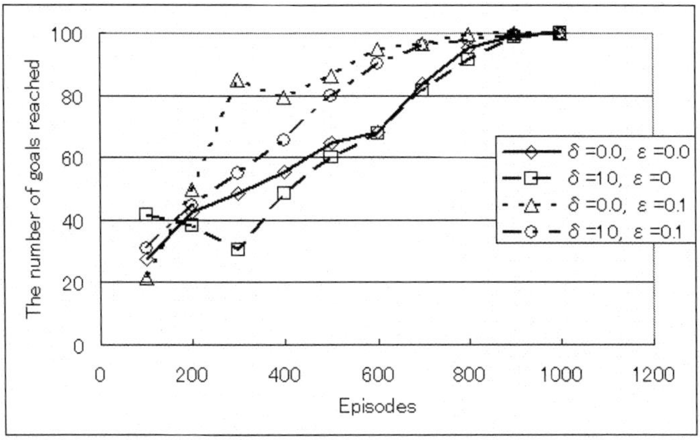

Fig. 4. The number of goals reached

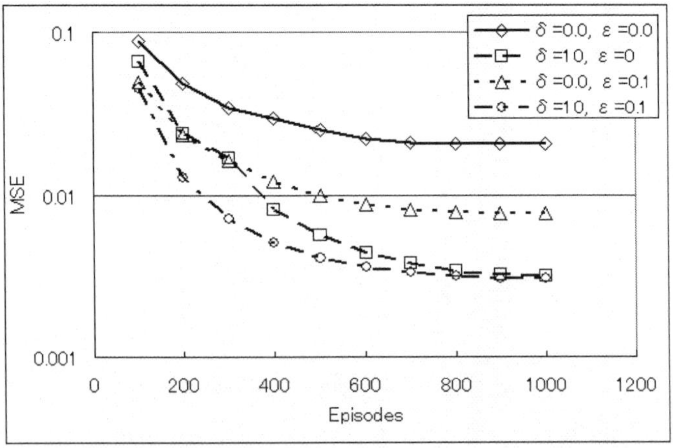

Fig. 5. MSE representing the knowledge on the environment

the agent knows the current surrounding, the larger negative reward it receives and the less action-value function becomes. We conjecture that a negative reward mechanism plays some role in learning of humans and animals.

Fig. 5 illustrates MSE representing one of the performance measures of a predictor. It demonstrates that the case with both positive and negative internal rewards gives the best performance.

4 Conclusions and Discussions

Experiments using internal rewards of curiosity and boredom well demonstrate that appropriate values of parameters simultaneously improve the performance of the predictor of the environment and increase the external rewards compared with the conventional reinforcement learning.

The role of curiosity and boredom might be similar to active learning in that data for training are selected to accelerate learning. The conventional active learning adopts a simple hypothesis on data selection. In contrast to this, the proposed method provides a detailed mechanism inspired by the characteristics of humans. The performance comparison is left for future study.

Developmental psychology uses two kinds of curiosity, diversive curiosity and specific curiosity. A typical example of the former is 2-3-year old infants having interests in many surrounding things. A typical example of the latter is 4-year-old infants having concentrated interests in small number of things and explore them in depth. Based on experimental results we conjecture that curiosity reward corresponds to specific curiosity, and boredom reward corresponds to diversive curiosity. Much more still remains to be done to mention the relationship with more confidence.

Acknowledgement

This research was partially supported by a COE program (center #J19) granted to Kyushu Institute of Technology by MEXT of Japan.

References

1. Sutton, R.S., Barto, A.G.: Reinforcement Learning. MIT Press, Cambridge (1998)
2. Shmidhuber, J.: Self-motivated development through rewards for predictor errors / improvements. In: Developmental Robotics, 2005 AAAI Spring Symposium (2005)
3. Oudeyer, P.-Y., Kaplan, F., Hafner, V.V.: Intrinsic Motivation Systems for Autonomous Mental Development. IEEE Trans. EC 11(2), 265–286 (2007)
4. Stout, A., Konidaris, G.D., Barto, A.G.: Intrinsically motivated reinforcement learning: A promising framework for developmental robot learning. In: Developmental Robotics AAAI Spring Symp. (2005)

Skill Transfer of a Mobile Robot Obtained by Reinforcement Learning to a Different Mobile Robot

Keiji Kamei[1,*] and Masumi Ishikawa[2]

[1] Department of Electrical, Electronics and Information Technology, Nishinippon Institute of Technology, 1-11, Niitsu, Kanda-machi, Fukuoka 800-0394, Japan
kamei@nishitech.ac.jp

[2] Department of Brain Science and Engineering, Graduate School of Life Science and Systems Engineering, Kyushu Institute of Technology, 2-4 Hibikino, Wakamatsu, Kitakyushu 808-0196, Japan
ishikawa@brain.kyutech.ac.jp

Abstract. Reinforcement learning (RL) is suitable for navigation of a mobile robot. We overcame some difficulties of RL which are large computational cost and determination of parameter values for RL with the help of a genetic algorithm (GA) and method of parameter prediction based on results of GA and complexity measure. As a result of these proposals, we succeeded in navigating the real robot practically. In our previous studies, we just one kind of mobile robot, which has three wheels. Our RL method can decrease the computational cost for learning of navigation and development of mobile robots, provided the skill obtained by RL for one mobile robot can be transferred to other mobile robots.

To verify the generalization capability of RL in navigation of a mobile robot, the present paper proposes to transfer the skill obtained by RL to a different kind of a mobile robot. We carried out the experiment and we succeeded in transferring the skill obtained by RL to a different mobile robot.

1 Introduction

Reinforcement learning (RL)[1] is suitable for navigation of a mobile robot. On the other hand, RL has some difficulties, i.e., large computational cost and determination of parameter values. To overcome these difficulties, we adopted following methods. Firstly, we introduced sensory information to improve its learning performance. Secondly, we utilized a genetic algorithm with inheritance for optimization of parameter values in RL. Thirdly, we predicted the values of parameters in RL for a novel environment by complexity measure of the environment. As a result of these proposals, we succeeded in navigating the real robot practically[2, 3, 4, 5, 6].

In our previous studies, we use just one kind of mobile robot, "*TRIPTERS mini*" which has three wheels. However, there are various kinds and types of mobile robots in the world, e.g., two wheels, four or more wheels and under water vehicles.

* Corresponding author.

A. Hanazawa et al. (Eds.): Brain-Inspired Info. Technology, SCI 266, pp. 57–62.
springerlink.com

Our RL method can decrease the computational cost for learning of navigation and development of mobile robots, provided the skill obtained by RL for one mobile robot can be transferred to other mobile robots.

To verify the generalization capability of RL in navigation of a mobile robot, the present paper proposes to transfer the skill obtained by RL using *TRIPTERS mini* to a different kind of a mobile robot. We carried out the experiment and we succeeded in transferring the skill obtained by RL to a different mobile robot.

2 Reinforcement Learning

We adopt the Q-learning, which is one of reinforcement learning(RL) methods. Q-learning estimates a Q-value, Q(s,a), as a function of a pair of a state and an action, which we think is suitable for a mobile robot. The conventional Q-learning iteratively updates a Q-value as,

$$Q(s,a) \leftarrow Q(s,a) + \alpha[r + \gamma \max_{a'} Q(s',a') - Q(s,a)] \quad (1)$$

where s is the current state, a is the corresponding action, s' is the next state, a' is the corresponding action, α is a learning rate, γ is a discount rate, and r is a reward or a penalty from the environment. A penalty, which is a negative reward, is also referred to as a reward for simplicity. The state, s, is composed of the location and orientation.

Major reason for slow learning in RL is that a mobile robot learns only at the current state based on a reward from the environment. To accelerate learning, we propose to directly reduce Q-values on a line segment between an obstacle and the current location of a mobile robot, in addition to the modification in Eq.(1). This proposal is called "modified Q-learning"[2].

3 Genetic Algorithm for Determination of Parameter Values for RL and Its Prediction Based on Complexity Measure

A GA is inspired by evolution of living things; search involves genetic operations such as selection, crossover and mutation in a probabilistic way frequently observed in the real evolutionary processes. Each individual has its chromosome and is evaluated by a fitness function. RL has various parameters such as a learning rate, a discount rate and rewards. The values of parameters in RL are not known in advance. We optimize the values of parameters in RL with the help of a GA. The computational cost for the optimization is very large in order to actually calculate the RL. To solve this difficulty, we adopt inheritance which is additional learning from final Q-values from previous generation[3, 4].

The proposal of GA with inheritance is drastically down of the computational cost. However, that had to optimize the parameter values in RL for every environment, which was impractical. We propose a hypothesis on the complexity of the environment: two major measures are the number of possible states and the ratio of the number of rotation actions to the total number of actions. The former means that the larger the number of possible states is, the more complex the environment becomes. The latter means that the larger the ratio is, the more complex the environment becomes. This proposal succeeded in predicting the optimal values of parameters in RL[5, 6].

4 Mobile Robots

We use two kind of mobile robots, two and three wheel mobile robot. Fig.1 shows those mobile robots.

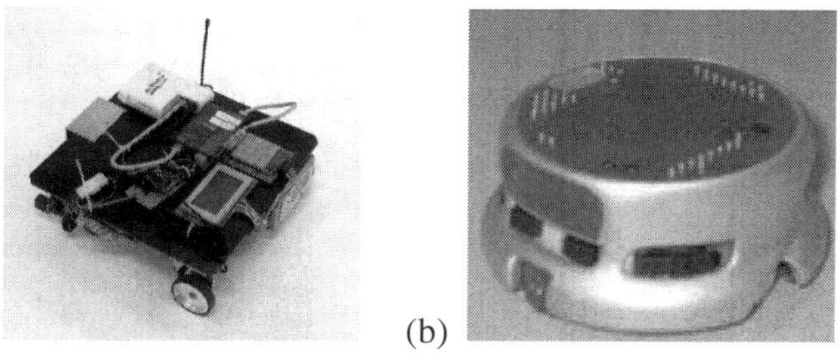

(a) (b)

Fig. 1. The mobile robots. (a)*TRIPTERS mini* (b)*Khepera II*

4.1 Three Wheel Mobile Robot, *TRIPTERS mini*

Fig.1(a) shows the three wheel mobile robot, *TRIPTERS mini*. *TRIPTERS mini* has one steering wheel and two independent driving wheels. The robot is not able to rotate on the spot because the axle between the 2 driving wheels. An action of turning actually moves slightly backward due to this property. The form of *TRIPTERS mini* is 30cm regular tetragon.

The *TRIPTERS mini* has 4 ultrasonic sensors and 4 infrared sensors. We use ultrasonic sensors, because of the ability of measuring the distance between the *TRIPTERS mini* and an obstacle. In addition, *TRIPTERS mini* has a rotary encoder in order to calculate odometry information.

4.2 Two Wheel Mobile Robot, *Khepera II*

Fig.1(b) shows the two wheel mobile robot, *Khepera II*. *Khepera II* has two independent driving wheels. Since the feature, it can turn without backward moving. *Khepera II* is 7cm circular form.

The *Khepera II* has 8 proximity sensors and 8 ambient sensors. However, we do not use these sensors. The *Khepera II* does not learn because of skill transfer from *TRIPTERS mini*. The *Khepera II* has rotary encoder same as *TRIPTERS mini*. This ability makes to calculate odometry information.

5 Skill Transfer; *TRIPTERS mini* to *Khepera II*

In our previous study, we developed a simulator for the *TRIPTERS mini*. It uses three primitive actions, i.e., moving forward by 10cm and turning right or left 10°.

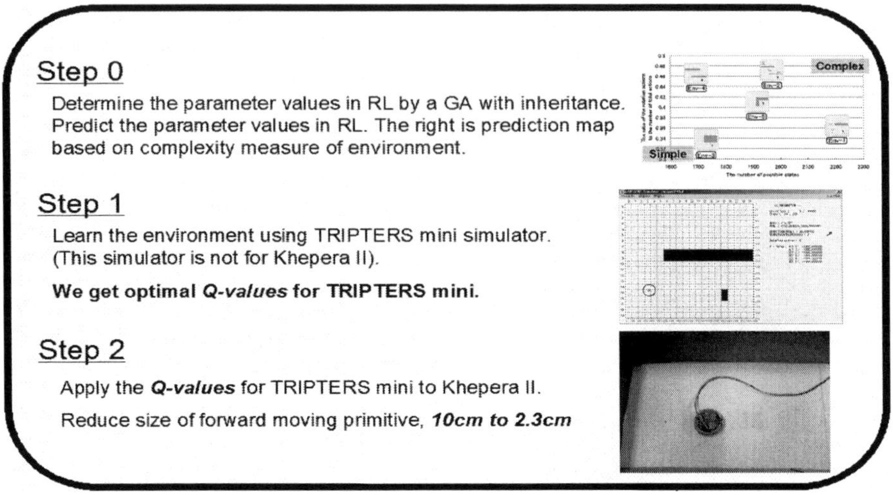

Fig. 2. Process of the experiment

A multiple of 10cm or 10° can easily be realized by sequence of the corresponding primitive. As these primitives smaller, the resulting path becomes more precise and the computational cost increase. Taking this trade off into account, we adopt above three primitive actions. We assume that *TRIPTERS mini* and *Khepera II* knows their position and orientation based on odometry information.

Fig.2 shows the process of this experiment. Initially, we use the simulator for learning an environment. We go on to apply the Q-values for *TRIPTERS mini* to the *Khepera II*. When the result of learning applies to it, we reduce size of forward moving primitive, 10*cm* to 2.3*cm*, because the *TRIPTERS mini* is bigger than the *Khepera II* about four times. On the other hand, the rotation action does not change.

6 Experimental Results

Fig.3(a) illustrates the result of RL using the TRIPTERS mini simulator and Fig.3(b) shows the environment for the *Khepera II*.

The big black object is an obstacle, the small one is the goal and the circles are trajectory of TRIPTERS mini on the simulation in Fig.3(a). The area of environment for *TRIPTERS mini* is 4*m* regular tetragon. On the other hand, the area of environment for *Khepera II* is 93*cm* regular tetragon because of the size of *Khepera II* is smaller than about 4 times. The numbers in Fig.3(b) are turning spot which correspond to the numbers in Fig.3(a).

Fig.4 shows the results of the experiment. The *Khepera II* goes a tad too far the turning spot except for turning spot 1. The distance of overshoot is about 1.5*cm*. The reasons of this result, the odometry errors increase on the turning spot. The turning action for *TRIPTERS mini* moves backward slightly, on the other hand, the *Khepera II* does not move backward. Moreover, the robot takes many turning

(a) 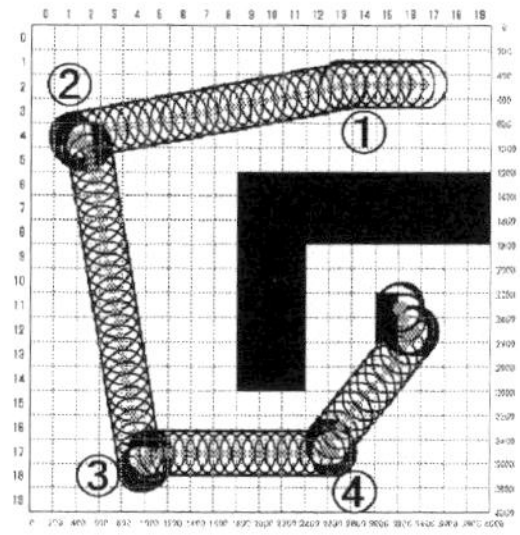

(b) 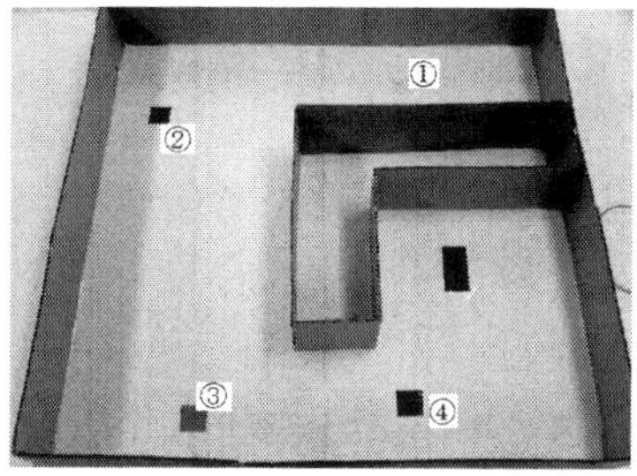

Fig. 3. (a)Result of RL using the simulator for *TRIPTERS mini* (b)The environment for *Khepera II*

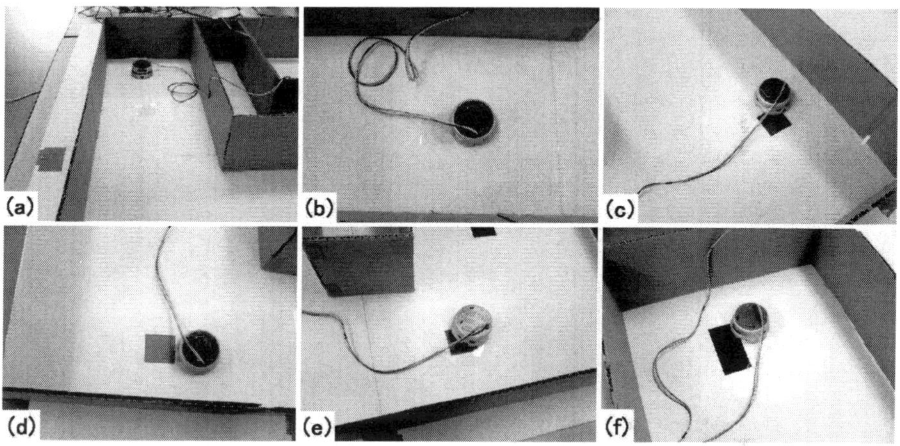

Fig. 4. Results of experiment. (a)Start (b)Spot 1 (c)Spot 2 (d)Spot 3 (e)Spot 4 (f)Goal

actions in the spot 2, 3 and 4. Fig.4(f) illustrates that the *Khepera II* reaches to the goal. It shows that the reaching coordinate is not so different from the result of simulation in Fig.3(a).

7 Conclusions

In this paper, we demonstrate the ability of generalization of RL. To verify, we apply the result of RL by *TRIPTERS mini* which has three wheels to *Khepera II* which has two wheels. The experimental result is that the *Khepera II* goes small detour, about 1.5cm. The *Khepera II*, however, reaches to neighbor of the goal regardless of the different motion for turning.

This experiment indicates that the optimal Q-values for one mobile robot is suboptimal Q-values for another mobile robot. This result shows that the skill obtained by RL for a mobile robot will be able to transfer other mobile robots.

Acknowledgements

This research was supported by the 21st Century Program, and by Grant-in-Aid for Scientific Research(C)(15500140) both from MEXT, Japan.

References

1. Sutton, R.S., Barto, A.G.: Reinforcement Learning. MIT Press, Cambridge (1998)
2. Kamei, K., Ishikawa, M.: More Effective Reinforcement Learning by Introducing Sensory Information. In: International Joint Conference on Neural Networks, pp. 3185–3188 (2004)
3. Kamei, K., Ishikawa, M.: Improvement of performance of reinforcement learning by introducing sensory information and a GA with inheritance. In: International Conference on Neural Information Processing 2005, pp. 1148–1153 (2005)
4. Kamei, K., Ishikawa, M.: Reduction of computational cost in optimization of parameter values in reinforcement learning by a genetic algorithm. In: Brain inspired-IT 2005. International Congress Series, vol. 1291, pp. 185–188 (2005)
5. Kamei, K., Ishikawa, M.: Dependency of Parameter Values in Reinforcement Learning for Navigation of a Mobile Robot on the Environment. Neural Information Processing - Letters and Reviews 10(7-9), 219–226 (July - September)
6. Kamei, K., Ishikawa, M.: Prediction of the optimal parameter values in reinforcement learning as a function of the environment. Brain inspired-IT 2006 Edited book (to appear)

Effect of Fitness Functions on the Performance of Evolutionary Particle Swarm Optimization

Hong Zhang and Masumi Ishikawa

Department of Brain Science and Engineering, Graduate School of Life Science and Systems Engineering, Kyushu Institute of Technology, 2-4 Hibikino, Wakamatsu, Kitakyushu 808-0196, Japan
{zhang,ishikawa}@brain.kyutech.ac.jp

Abstract. This paper presents an Evolutionary Particle Swarm Optimization (EPSO) for PSO model selection. It provides a new paradigm of meta-optimization that systematically estimates appropriate values of parameters in PSO model for efficiently solving various optimization problems. In order to further investigate the characteristics, i.e., exploitation and exploration in search, of the PSO model optimized by EPSO, we propose to use two fitness functions in EPSO, which are a temporally cumulative fitness of the best particle and a temporally cumulative fitness of the entire swarm for designing PSO models. Applications of the proposed method to some benchmark optimization problems well demonstrate its effectiveness. Our experimental results indicate that the former fitness function can generate a PSO model with higher fitness, and the latter can generate a PSO model with faster convergence.

1 Introduction

Particle Swarm Optimization (PSO) is a stochastic and population-based adaptive optimization algorithm proposed by J. Kennedy and R.C. Eberhart [3,4]. During the last decade, many variants of PSO were proposed, and successfully applied to a lot of research and application disciplines [5,6,9].

Although the mechanism of the original PSO is very simple, however, how to appropriately determine the values of parameters in PSO for efficiently solving various optimization problems is still attractive challenging problem. Many researchers paid much attention to the above problem, and proposed several algorithms for dealing with it. Specifically, one approach is to try different values of parameters to find a proper parameter set of PSO for handling many kinds of optimization problems reasonably well [1,2]. Another approach is Composite PSO (CPSO) [8], in which the Differential Evolution (DE) algorithm [10] copes with the PSO heuristics in parallel during optimization. It is to be noted that an instantaneous fitness for evaluating the performance of PSO.

A. Hanazawa et al. (Eds.): Brain-Inspired Info. Technology, SCI 266, pp. 63–68.
springerlink.com

For systematically estimating a PSO model, Meissner et al. proposed an Optimized Particle Swarm Optimization (OPSO) method for meta-optimization of PSO heuristics [7]. Zhang et al. proposed an Evolutionary Particle Swarm Optimization (EPSO) method, in which the Genetic Algorithm (GA) is used for estimating a proper parameter set in PSO by a temporally cumulative fitness for evaluating the performance of PSO [12]. Since the probability of the accident in estimating PSO model can be greatly reduced by the temporally cumulative fitness, EPSO provides a powerful methodology for PSO model selection.

In order to further investigate the characteristics, i.e., exploitation and exploration in search, of the PSO model optimized by EPSO, here we propose to use two fitness functions in EPSO, which are a temporally cumulative fitness of the best particle and a temporally cumulative fitness of the entire swarm for systematically designing and selecting PSO models. The former was already used in our previous study [12], the latter is a new proposal that emphasizes the behavior of entire particles based on majority decision in social behavior.

2 Proposed Method

2.1 About PSO

The original PSO is modeled by particles with position, $\boldsymbol{x}$, and velocity, $\boldsymbol{v}$, as follows,

$$\begin{cases} \boldsymbol{x}_{k+1}^i = \boldsymbol{x}_k^i + \boldsymbol{v}_{k+1}^i, \\ \boldsymbol{v}_{k+1}^i = c_0\boldsymbol{v}_k^i + c_1\boldsymbol{r}_1(\boldsymbol{x}_l^i - \boldsymbol{x}_k^i) + c_2\boldsymbol{r}_2(\boldsymbol{x}_g - \boldsymbol{x}_k^i), \end{cases}$$

where i is the index of particle, k is time-step, c_0 is an inertial factor, c_1 is an individual confidence factor, c_2 is a swarm confidence factor, and $\boldsymbol{r}_1, \boldsymbol{r}_2 \in [0, 1]$ are random values, $\boldsymbol{x}_l^i (= \arg\max_{k=1}^{K}\{g(\boldsymbol{x}_k^i)\})$ is the best position for the particle $\boldsymbol{x}^i$, and $\boldsymbol{x}_g (= \arg\max_{i=1}^{P}\{g(\boldsymbol{x}_l^i)\}$; P: the number of the swarm particles) is the best position for the swarm.

It is well known that although the mechanism of the original PSO is simple, it can provide better results compared with other methods such as machine learning, neural network learning and so on. However, determining appropriate values of parameters in PSO, i.e., c_0, c_1, and c_2, for efficiently solving a given optimization problem is an important point (called meta-optimization).

2.2 EPSO

EPSO provides a new paradigm of meta-optimization that systematically estimates a proper parameter set in PSO for finding an optimal solution to a given optimization problem. Specifically, the procedure of EPSO is composed of two parts (see Fig. 1): One is a Real-coded Genetic Algorithm with Elitism strategy (RGA/E) [11]. While the PSO finds a solution to a given optimization problem,

RGA/E is used for optimizing the values of parameters in PSO with online evolutionary computation. The other is PSO. The PSO with the parameter values provided by RGA/E efficiently solve a given optimization problem.

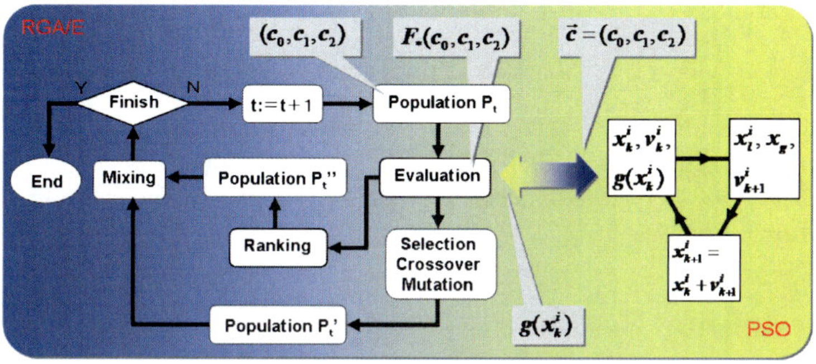

Fig. 1. The flowchart of EPSO

For investigating the characteristics of the PSO model optimized by EPSO, the following two fitness functions are applied in EPSO. The one is a temporally cumulative fitness, F_1, of the best particle, and the other is a temporally cumulative fitness, F_2, of the entire swarm, which are defined by

$$F_1(c_0,c_1,c_2)=\sum_{k=1}^{K} g(x_k^b)\Big|_{c_0,c_1,c_2} \quad ; \quad F_2(c_0,c_1,c_2)=\sum_{k=1}^{K} \overline{g}_k\Big|_{c_0,c_1,c_2} ,$$

where $g(x_k^b)$ is the fitness value of the best particle, $\overline{g}_k$ is the average fitness value of the swarm particles, and K is the number of the maximum iteration in PSO.

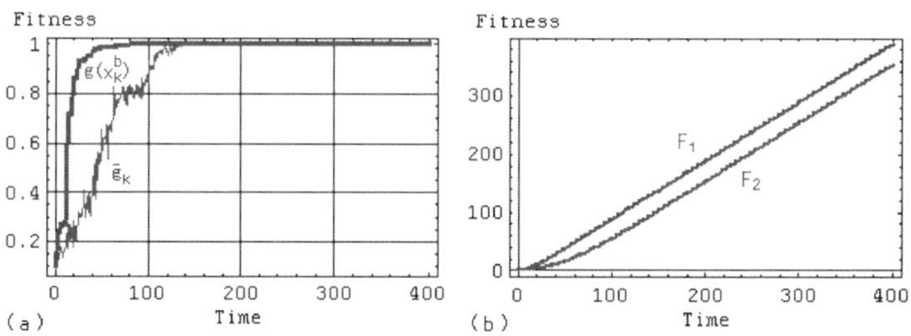

Fig. 2. Comparison of variations with two temporally fitness functions, F_1 and F_2

Fig. 2 illustrates the relationship between two fitness functions, F_1 and F_2, for evaluating the performance of PSO model. We can clearly see that although two instantaneous fitness functions, $g(x_k^b)$ and $\overline{g}_k$, have different characteristics, the corresponding temporally cumulative fitness functions, F_1 and F_2, almost linear increase with very small fluctuation. It empirically suggests that the fitness functions, F_1 and F_2, are suitable to PSO model selection for evaluating the performance of PSO model.

In order to investigate the convergence property of the PSO model optimized by EPSO with the above fitness functions, we define the maximum time-step, k_{max}, as a criterion index by

$$\forall \boldsymbol{k} \geq \boldsymbol{k}_{\max}, \ g(\boldsymbol{x}_k^b) - \bar{g}_k \leq \tau .$$

Here, τ is a positive parameter. Based on the fundament on the smaller the maximum time-step, k_{max}, is, the faster the convergence speed of the swarm is, we can identify the convergence property of PSO model for solving a given optimization problem.

3 Experimental Results

Computer experiments are carried out by EPSO with two fitness function, F_1 and F_2, for solving the following benchmark optimization problems in Fig. 3. Parameters in PSO are estimated under the condition that c_0, c_1, and c_2 are not negative.

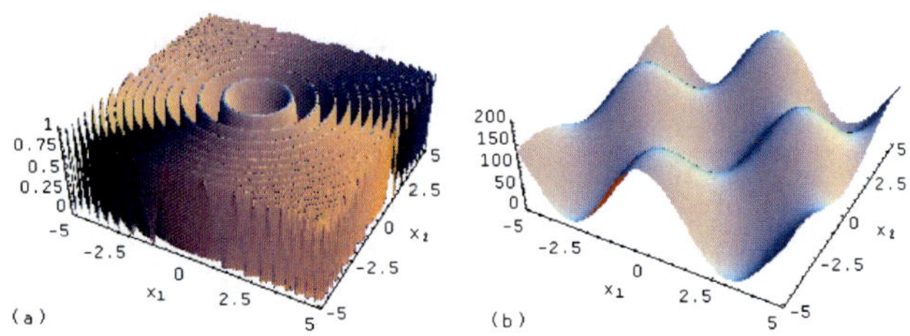

Fig. 3. Optimization problems. (a) Schaffer's F6 problem, (b) Griewank problem.

Table 1 shows the resulting parameter values in PSO by EPSO for two fitness functions, F_1 and F_2, and the corresponding average fitness of the best particle, $\bar{g}(\boldsymbol{x}_g)$, and the maximum time-step, k_{max}. By comparison with their data, we observed that the average values estimated by the fitness function, F_1, is in most cases larger than that by the fitness function F_2, and the convergence time-step by the fitness function, F_2, is shorter than that by the fitness function F_1.

Table 1. The experimental results (P=10, K=400, τ =0.03 for 20 trials)

Problem	Fitness	c_0	c_1	c_2	$\bar{g}(\boldsymbol{x}_g)$	k_{max}
Schaffer	F_1	1.21±0.8	3.17±2.0	2.34±0.7	0.998±0.006	400
	F_2	0.03±0.0	0.52±0.6	1.09±0.4	0.997±0.004	8.60
Griewank	F_1	0.21±0.2	0.91±0.8	2.93±3.9	0.994±0.007	34.0
	F_2	0.27±0.0	1.13±1.1	1.34±0.1	0.993±0.009	12.0

For the sake of visualization to PSO search, Fig. 4 illustrates the distribution of whole particles in the search space for solving the Schaffer's F6 problem. We observed that the search behavior and convergence property of the PSO models optimized by the fitness function, F_1 and F_2, are quite different. Although both of the resulting PSO models all can solve the optimization problem, from the viewpoint in the exploitation and exploration, the PSO model optimized by the fitness function, F_1, has unique performance than that by the fitness function, F_2.

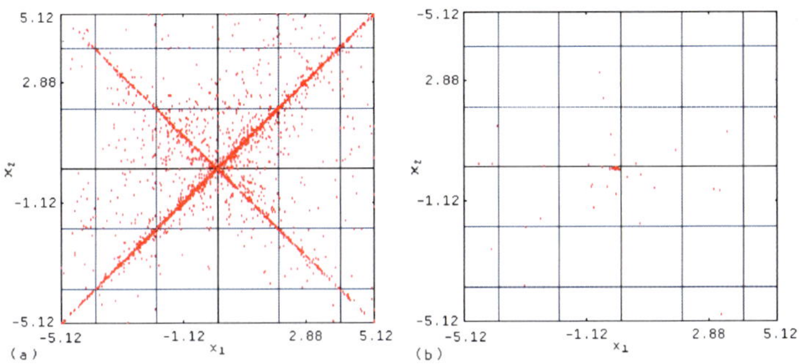

Fig. 4. Distribution of positions of whole particles. (a) Fitness function F_1. (b) Fitness Function F_2.

4 Conclusions and Discussions

We have proposed Evolutionary Particle Swarm Optimization, EPSO, and used two fitness functions, i.e., a temporally cumulative fitness of the best particle, F_1, and a temporally cumulative fitness of the entire swarm, F_2, in EPSO for PSO model selection. Our experimental results indicated that the former could generate a PSO model with higher fitness, and the latter could generate a PSO model with faster convergence. Since the PSO model optimized by the fitness function, F_1, has superior performance in balancing between exploitation and exploration in PSO search, it is recommended to use it to design PSO models for efficiently solving various optimization problems.

Even though better results have been obtained, only small-scale benchmark optimization problems were carried out. It is left for further study to apply the proposed method to high dimensional and more complex optimization problems in the real-world.

Acknowledgement

This research was supported by the 21st century COE (Center of Excellence) program (#J19) granted to Kyushu Institute of Technology form the Ministry of Education, Culture, Sports, Science and Technology (MEXT), Japan.

References

1. Beielstein, T., Parsoploulos, K.E., Vrahatis, M.N.: Tuning PSO Parameters Through Sensitivity Analysis. Technical Report of the Collaborative Research Center 531 Computational Intelligence CI-124/02, University of Dortmund (2002)
2. Carlisle, A., Dozier, G.: An Off-The-Shelf PSO. In: Proceedings of the Workshop on Particle Swarm Optimization, Indianapolis, pp. 1–6 (2001)
3. Eberhart, R.C., Kennedy, J.: A new optimizer using particle theory. In: Proceedings of the sixth International Symposium on Micro Machine and Human Science (1995), doi:10.1109/MHS. 1995.494215
4. Kennedy, J., Eberhart, R.C.: Particle swarm optimization. In: Proceedings of IEEE International Conference on Neural Networks (1995), doi:10.1109/ICNN.1995.488968
5. Kennedy, J., Eberhart, R.C., Shi, Y.: Swarm Intelligence. Morgan Kaufmann Publishers, CA (2001)
6. Kennedy, J.: In Search of the Essential Particle Swarm. In: Proceedings of the 2006 IEEE Congress on Evolutionary Computations, Vancouver, BC, Canada, July 16-21, pp. 6158–6165 (2006)
7. Meissner, M., Schmuker, M., Schneider, G.: Optimized Particle Optimization (OPSO) and its application to artificial neural network training. BMC Bioinformatics 7, 125 (2006)
8. Parsopoulos, K.E., Vrahatis, M.N.: Recent approaches to global optimization problems through Particle Swarm Optimization. Natural Computing 1, 235–306 (2002)
9. Reyes-Sierra, M., Coello, C.A.C.: Multi-Objective Particle Swarm Optimizers: A Survey of the State-of-the-Art. International Journal of Computational Intelligence Research 2(3), 287–308 (2006)
10. Storn, R., Price, K.: Differential evolution – a simple and efficient heuristic for global optimization over continuous space. Journal of Global Optimization 11, 341–359 (1997)
11. Zhang, H., Ishikawa, M.: A Hybrid Real-Coded Genetic Algorithm with Local Search. In: Proceedings of the 12th International Conference on Neural Information Processing (ICONIP 2005), Taipei, Taiwan, R.O.C, October 30-November 2, pp. 732–737 (2005)
12. Zhang, H., Ishikawa, M.: Evolutionary Particle Swarm Optimization (EPSO) – Estimation of Optimal PSO Parameters by GA. In: Proceedings of The IAENG International MultiConference of Engineers and Computer Scientists (IMECS 2007), Newswood Limited, 1, Hong Kong, China, March 21-23, pp. 13–18 (2007)

Constraint Satisfaction with Neural Network for Trajectory Planning

Shinji Nagano, Yoichi Fujimoto, and Masahiro Nagamatu

Department of Brain Science and Engineering, Graduate School of Life Science and Systems Engineering, Kyushu Institute of Technology, 2-4 Hibikino, Wakamatsu, Kitakyushu 808-0196, Japan
{nagano-shinji,fujimoto-yoichi}@edu.brain.kyutech.ac.jp, nagamatu@brain.kyutech.ac.jp

Abstract. The trajectory planning is generally an important and difficult problem to solve when robots are required to do some actions. Robots should move in an efficient way satisfying several constraints imposed on them. The constraints include dynamics of the robots, the limitations of values of state and control variables, e.g. imitations of angles and torques of joints, obstacle avoidance and achievement of required tasks. Our proposed method is based on constraint satisfaction. By using this, the constraints mentioned above can be satisfied effectively and flexibly.

1 Introduction

When robots are required to do some actions, the trajectory planning is necessary. For the trajectory planning the following constraints must be considered: geometric constraints (e.g. obstacle avoidance, limitations about configurations, etc) and kinematic and dynamic constraints (e.g. nonholonomic constraints, limitations about velocities and accelerations, limitations about forces exerted, etc). When the world in which robots work is very large and complicated, the trajectory planning becomes a very difficult problem.

Many methods are already proposed for the trajectory planning. Shin et al. proposed the use of dynamic programming to solve the optimization problem [2]. This method is not suitable for application to a system that has a large state space.

Basically it can be considered as an inverse problem to obtain controls to achieve the given tasks efficiently under several constraints. Early studies consider only geometric constraints. Kinematic and dynamic constraints are then considered to apply it to more practical problems. Our proposed method considers both geometric constraints and kinematical and dynamical constraints. It is based on constraint satisfaction. By using this, the constraints mentioned above can be satisfied effectively and flexibly.

A. Hanazawa et al. (Eds.): Brain-Inspired Info. Technology, SCI 266, pp. 69–72.
springerlink.com

2 Proposed Method

The problem solved by our method is described as follows:

Given: An initial state, a target state (or conditions which must be satisfied at the target state), and constraints for state and control variables.

Objective: Determine controls and time durations which lead the initial state to the target state satisfying all constraints.

The overview of our method is described as follows:

1) Learning of the forward model.
2) Constructing cascaded forward model.
3) Considering constraints.
4) Solving the CSP and finding the trajectory planning.

2.1 Learning of the Forward Model

The forward model of the dynamics of the robot is learnt and represented by a multi-layer perceptron (MLP) (Fig.1). The inputs of the MLP are a state of the robot, a control and time duration. The output is a state reached after the control. By using MLP, even if mathematical dynamics is unknown, the forward model can be learnt from examples of the input-output pairs of the robot.

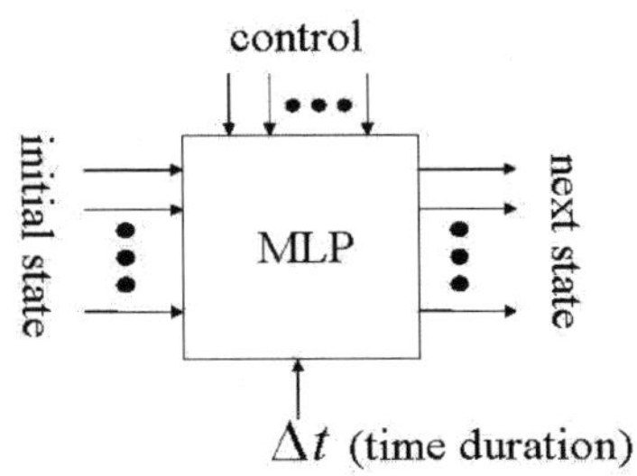

Fig. 1. Multi-layer perceptron used in the proposed method

2.2 Constructing Cascaded Forward Model

The copies of the MLP are cascaded sequentially to construct a large neural network (Fig.2). Trajectory planning is done by solving the inverse problem and by

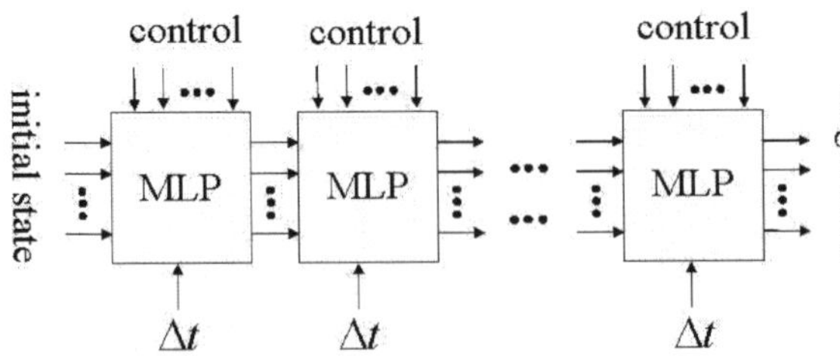

Fig. 2. Cascaded multi-layer perceptrons.

determining the controls and time durations. The number of MLPs in the network is determined by the total length of the controls required to reach the target state from the initial state. Generally, the length of the controls is not known in advance, hence, in our methods the number of MLPs can be determined dynamically.

2.3 Considering Constraints

After constructing cascaded forward model, constraints imposed on the robot are considered (Fig.3). The constraints are:

1) The output of cascaded multi-layer perceptrons must equal to the target state.
2) The limitations of the control variables.
3) The limitations of the values of the state.

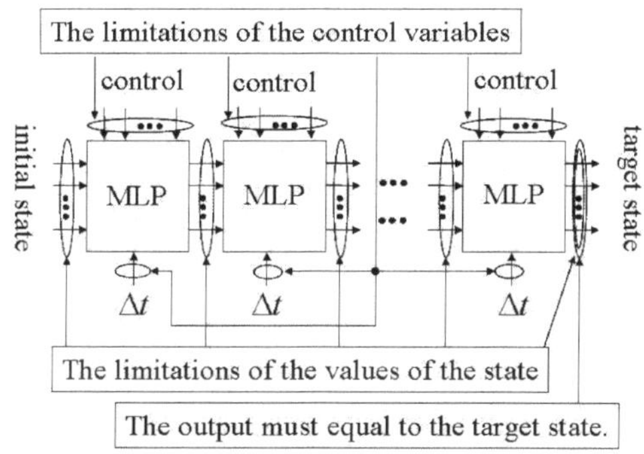

Fig. 3. Constraints

2.4 Solving the CSP and Finding the Trajectory Planning

The trajectory planning is a kind of inverse problem in which controls are determined to obtain the required trajectory. We consider this inverse problem as a constraint satisfaction problem (CSP). The CSP is solved by a neural network called LPPH (Lagrange Programming neural network with Polarized High-order connections) [1]. The LPPH is a neural network proposed for solving CSP.

3 Experiment Results

We apply our method to the trajectory planning of a 3-link robot-arm which is implemented in a simulator. The second and third joints can exert torque, while the first cannot. Therefore it is an example of underactuated robot. The trajectory planning of this robot-arm does not have so many constraints, but for underactuated robot it is known to be hard. So we think, for the first step, this robot-arm is suitable to check the effectiveness of our method. In the experiment we solved a trajectory planning of recovering the robot-arm from a tottering state to a stable inverted state by exerting torque at the second and third joints. In this case, each state in Fig.1 consists of angles (0 rad~2π rad) and angular velocities (−20 rad/s~20 rad/s) of joints. Each control consists of torques (−16.1 N·m~16.1 N·m) of joints. Each time duration is 0 s~0.1 s. The constraints are 1) the output of the last MLP must equal to the target state, 2) the values of state and control variables must be within the limitations. The LPPH solves the constraint satisfaction problem by changing values of the variables and weights of the constraints according to the LPPH dynamics. The trajectory obtained by the experiment is shown in Fig.4. The initial and target states are colored by black. In the experiments we succeeded in recovering the robot-arm by using cascaded 10 MLPs.

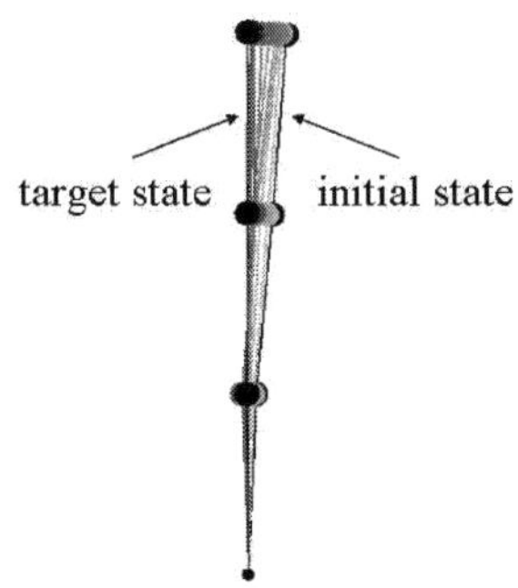

Fig. 4. Result of trajectory planning of three link robot-arm

4 Discussions

We can solve trajectory planning for simple underactuated robot. We want to apply our method to robots which has more complex dynamics and constraints. The CPU time for obtaining the result of Fig.2 is about 30 sec on PENTIUM III 1.5GHz. It seems to be hard to use this method for online usage. We think we can improve our method, however, we also think the best way is to combine our method with reinforcement learning. By making successful trajectories by out method, and by feeding them to the reinforcement learning, we think we can drastically speedup the learning process of the reinforcement learning.

References

1. Nakano, T., Nagamatu, M.: A continuous valued neural network with a new evaluation function of degree of unsatisfaction for solving CSP. IEICE, E89-D, 4 (2006)
2. Shin, K.G., Macky, D.N.: A dynamic programming approach to trajectory planning of robotic manipulators. IEEE Trans. Autom. Control AC-31(6), 450–501 (1986)

Acquisition and Extinction of Avoidance Response by Social Interaction in Rats

Akira Masuda*, Kimiya Narikiyo, Noboru Shiota, and Shuji Aou

Department of Brain Science and Engineering, Graduate School of Life Science and Systems Engineering, Kyushu Institute of Technology, 2-4 Hibikino, Wakamatsu, Kitakyushu 808-0196, Japan
masuda-akira@edu.brain.kyutech.ac.jp

Abstract. In rodents, many forms of behavior (avoidance, food choice, exploring, etc.) have been shown to be transmitted through social interaction. Previous psychological studies have shown that acquisition of avoidance behavior is under the influence of social interaction but detail analyses have not been done. In the present study, we studied how learned avoidance behavior is influenced by interacting with other rats which have different experiences to foot shock using passive avoidance paradigm. Rats received electrical shocks in the dark compartment and avoided entering there were used as subjects. In Experiment 1, the latency to enter the dark compartment was measured under three different conditions; without any partner (ALONE), with a shock-received partner (wSHOCK) and with a unshocked-naïve partner (wNAIVE). The latencies of wSHOCK and wNAIVE to enter the dark compartment were shorter than that of ALONE. In Experiment 2, the effects of other rats' behavior on reinstate of avoidance were examined. The subject rats used in the Experiment I were put in the dark compartment without electrical foot shock and habituated to the compartment. Then the latencies to enter the dark compartment were measured in absence or presence of demonstrator rat which received electric foot shock in front of the subject rats. The demonstration of receiving foot shock by other rats enhanced avoidance responses. These results suggest that learned avoidance may be modified by partners' behaviors in reciprocal fashion.

1 Introduction

Many social animals adapt to their environments by sharing information among the conspecifics (fish [5], birds [1,7], monkeys [6]). Rodents, commonly used experimental animals, also have been investigated in this perspective. For example, the previous studies have shown that food choices of rats are influenced by which food was eaten by other rats [2, 3]. Avoidance to something dangerous is thought to be one of the most important behavioral choices to survive. Some studies showed, however, avoidance behavior is not easily transmitted by social interaction [4, 8]. One possible reason of such results may be because those studies used naïve rats as subjects to test how rats were influenced by the other rats. It is not

* Corresponding author.

A. Hanazawa et al. (Eds.): Brain-Inspired Info. Technology, SCI 266, pp. 73–78.
springerlink.com

well understood whether learned avoidance can be influenced by other rats. In the present study, we focused on avoidance-learned rats and evaluated effects of social interaction on learned avoidance behavior.

2 Materials and Methods

2.1 Animals

Thirty-two Wistar male rats aged 8 weeks were purchased from Kyudo co. were used. They were freely given food and water, and housed in two per cage for one week before the start of the experiments. Housing conditions were thermostatically-controlled at 22-24 °C with a light/dark cycle (lights on: 08:00—20:00). The experiments were performed under the control of the Ethics Committee of Animal Care and Experimentation in accordance with the Guiding Principles for Animal Care Experimentation, Kyushu Institute of Technology, Japan, and with the Japanese Law for Animal Welfare and Care.

2.2 Experimental Setup

The experiments took place in a grid floored chamber which consisted of two compartments; the light compartment (D25 cm×W25 cm×H27 cm) and the dark compartment (D30 cm×W30 cm×H30 cm). The two compartments were divided by a sliding door. Electric shocks are delivered to grid floor of dark compartment by a shock generator (SGS-002, Muromachi Kikai Co., Ltd., Tokyo, Japan). In Experiment 2, removable partition was used to prevent animals from moving.

2.3 Training (Passive Avoidance Test)

Before the Experiment 1 and 2, subjects were trained as follows. On the first day of the session, all animals were placed in the light compartment for 1 min for habituation individually. Then sliding door was raised and the latency to enter the dark compartment was recorded. On the second day, after the same procedure as the first day, the electrical shocks (0.5 mA, 5 sec) were applied to 24 rats in the dark compartment. Eight rats which did not receive electrical shocks were used as naïve rats. The experimental apparatus was cleaned with alcohol to remove odors before testing the next subject. On the third day, the latencies of each animal to entering the dark compartment were measured.

2.4 Experiment 1: Shock (-) Conditions

Animals were tested under three different conditions: i) alone (ALONE), ii) coupled with naive rats (wNAIVE), iii) coupled with shocked rats (wSHOCK).On the day following the training session (day 4), each subject was placed in the light compartment with or without other rat for 1 min. After the interval, the sliding door was raised and then the latencies to entering the dark compartment were measured with the cut-off time of 15 min. This experiment was performed under no electric shock condition.

2.5 Experiment 2: Shock (+) Conditions

The day next to Experiment 1 (day 5), subjects were put in the experimental apparatus individually and habituated for 20 min. On the second day of Experiment 2 (day 6), 30 min before the test trial, each animal was placed in the light compartment and then the latencies to entering the dark compartment were measured with cut-off time of 5 min. The animals which stayed in the light compartments over the cut-off time were precluded and avoidance behaviors of 11 animals were evaluated. Thirty minutes after the pre-test trial, each subject was placed in the light compartment with other rat for 1 min. After the interval, the sliding door was raised and then electrical shocks (0.5 mA) were induced to the other rats. The other rats always entered the dark compartment earlier than subjects because of the partition. After the additional interval (30 sec), the partition was removed. Then the latencies to enter the dark compartment were measured (cut-off time 15 min).

2.6 Data Analysis

In Experiment 1, statistical significance of the difference among the groups of learned animal was assessed using the Steel-Dwass multiple comparison test due to the cut-off time. In Experiment 2, statistical significance of difference between without demonstrator rats and with demonstrator rats was evaluated using the Wilcoxon signed rank test. The criterion for statistical significance was $P < 0.05$ (two-tailed).

3 Results

3.1 Experiment 1

First, we confirmed that the all foot-shocked subjects learned avoidance response by measuring latencies to enter the shock compartment. All subjects did not enter the dark compartment within 5 min (mean latency = 1127 sec, SEM = 31.9 sec). Second, we tested whether the learned avoidance would be modified by interaction with the other rats. The latency to enter the dark compartment was significantly shorter in wNAIVE subjects than both in ALONE subjects ($P<0.01$) and wSHOCK subjects ($P<0.05$). Interestingly, it was not significant difference, but the latency of wSHOCK was also suppressed rather than facilitated comparing to the ALONE ($P=0.1$) (Fig. 1).

3.2 Experiment 2

We used the same subject animals for evaluation of facilitating effects of social transmission because the avoidance behavior of avoider rats during this session can vary by individual variation. Before the test, we examined initial avoidance level measuring the latency of shocked rats under no socially interactive conditions, and 11 animals that entered the dark compartment within 5 min (cut-off time) were used. After 30 min interval, avoider subjects were placed in experimental setting again, and interacted with be-in-shock-demonstrators. Normalized

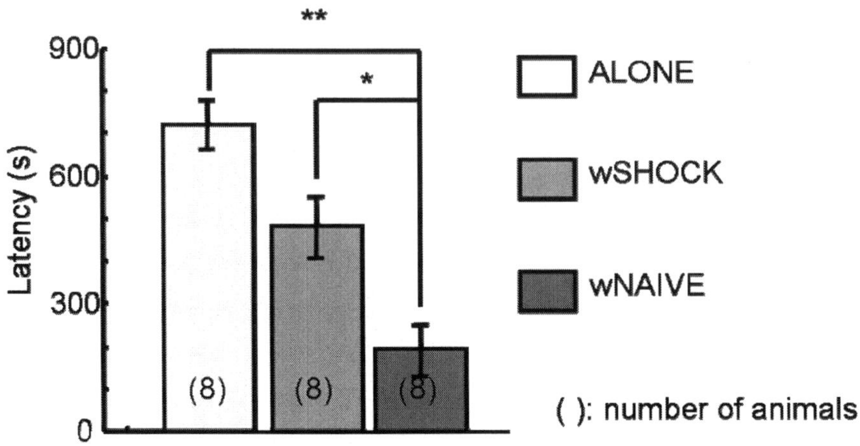

Fig. 1. Suppressive effects on avoidance responses by social interaction. Avoidance-learned rats with naïve partners (wNAIVE) entered the dark compartment significantly earlier (*: P<0.05, **: P<0.01, Steel-Dwass test) than those with shocked-partners (wSHOCK) or those without any partners (ALONE).

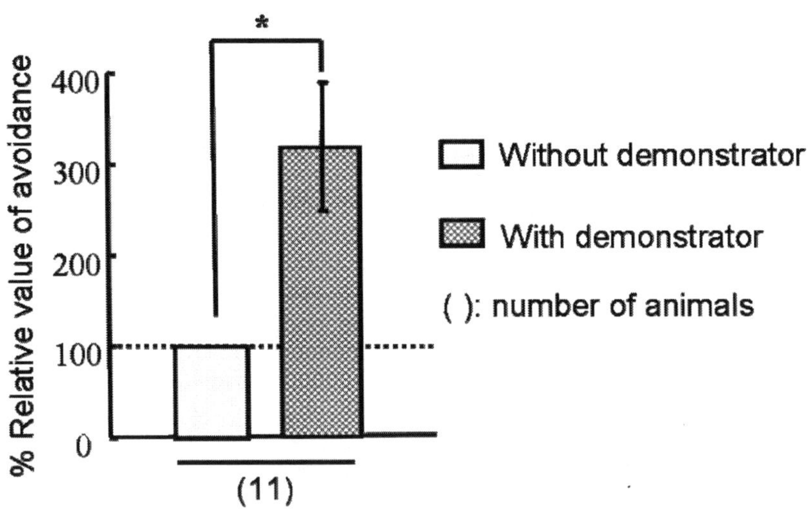

Fig. 2. Facilitating effect on avoidance behavior by social interaction. After foot shock stimuli to the demonstrator (with demonstrator), latency of avoidance-learned rats tended to be slower than previous latency (without demonstrators) (*: P<0.05, Wilcoxon test). Normalized latencies of animals with shocked demonstrator data are shown as average ±SEM.

latency of the avoidance-learned rats was significantly increased by the interaction (P<0.05). Demonstrating shocked behavior to the avoider facilitated avoidance responses (Fig. 2).

4 Discussion

On the shock (-) conditions, the partial extinction of avoidance responses were found in avoidance-learned rats by the other rats. Especially, this suppressive effect was greater in wNAIVE rats than in wSHOCK rats. Although it was not significantly different, the avoidance responses of wSHOCK were less than those of Alone subjects. This can be explained theoretically by a probabilistic difference. In the condition of two learned rats, each of them enters dark compartment more likely in a time-limit than single. Even if there was not suppression on avoidance response of individual subject in the condition, the latency could be shortened. Another possibility is an effect by the population density. In the condition, two learned rats were in the limited space. Because they were housed in 2 per cage, not in isolation, their aggressiveness was not strong, but the high population density could motivate to transfer the compartment. These results suggest that the avoidance responses can be inhibited by social interaction on no-shock environments.

On the shock (+) conditions, interaction with the other shocked rats enhanced the avoidance responses in avoidance-learned rats. Together with the result of previous studies, naïve rats were not influenced by the shocked demonstrators [4, 8]. Social transmission of avoidance learning may require sharing same experiences or memories together in rats, because avoiders had experienced passive avoidance learning previously. In other word, the avoidance could be transmitted socially in experience-dependent fashion. From these results, social transmission of avoidance behavior may require to share same experiences or memories together in rats. This view might provide new points to ponder to studies concerning social transmission in laboratory.

In conclusion, the efficacy of social transmission in avoidance-learned rats is greater than that in naive rats. Since social transmission of learning processes have been widely observed in various species, understanding of the neural mechanism of social transmission may help to reveal collective intelligence.

Acknowledgement

This research was supported by 21st century COE program, and Grants-in Aid for Scientific Research from the Ministry of Education, Science & Culture of Japan.

References

1. Burt de perera, T., Guilford, T.: The social transmission of spatial information in homing pigeons. Anim. Behav. 57, 715–719 (1999)
2. Galef Jr., B.G., Whiskin, E.E.: Socially transmitted food preferences can be used to study long-term memory in rats. Learn. Behav. 3, 160–164 (2003)
3. Galef Jr., B.G., White, D.J.: Socially acquired information reduces rats' latencies to find food. Anim. Behav. 54, 705–714 (1997)
4. Galef Jr., B.G., McQuoid, L.M., Whiskin, E.E.: Further evidence that Norway rats do not socially transmit learned aversions to toxic baits. Anim. Learn. Behav. 18, 199–205 (1990)

5. Hall, D., Suboski, M.D.: Visual and olfactory stimuli in learned release of alarm reactions by zebra danio fish (Brachydanio rerio). Neurobiol. Learn. Mem. 63, 229–240 (1995)
6. Hikami, K., Hasegawa, Y., Matsuzawa, T.: Social transmission of food preferences in Japanese monkeys (Macaca fuscata) after mere exposure or aversion training. J. Comp. Psychol. 104, 233–237 (1990)
7. Strasser, R., Bingman, V.P.: The effects of hippocampal lesions in homing pigeons on a one-trial food association task. J. Comp. Physiol. 185, 583–590 (1999)
8. White, D.J., Galef Jr., B.J.: Social influence on avoidance of dangerous stimuli by rats. Anim. Learn. Behav. 26, 433–438 (1998)

Positive and Negative Effects of Environmental Chemicals on Brain Function in Rodents

Tomoko Tsuruoka[1,*], Tetsuya Fujimoto[1], Noboru Shiota[1], Makoto Monda[1], Yukiko Fueta[2], Toru Ishidao[2], Hajime Hori[2], and Shuji Aou[1]

[1] Department of Brain Science and Engineering, Graduate School of Life Science and Systems Engineering, Kyushu Institute of Technology, 2-4 Hibikino, Wakamatsu, Kitakyushu 808-0196, Japan
tsuruoka-tomoko@edu.brain.kyutech.ac.jp
[2] Department of Environmental Management, School of Health Science, University of Occupational and Environmental Health, 1-1 Iseigaoka, Kitakyushu 807-8555, Japan

Abstract. We investigated the effect of environmental chemicals on exploratory behavior and emotional behavior in Wistar rats, using behavioral task such as open-field test, elevated plus maze test and forced swimming test. We used green odor and jasmine tea odor as natural environmental chemicals and 1-bromopropane (1-BP) and bisphenol-A (BPA) as artificial environmental chemicals. Green odor did not affect sensory-motor system but attenuated swimming-induced decrease in activity suggesting its anti-fatigue effect. Jasmine tea odor increased exploratory behavior in normal physiological condition. On the other hand, prenatal exposure 1-BP and BPA impaired sexual differentiation of exploratory behavior and BPA increases depression-like behavior. The results suggest that animal behavior are highly sensitive to external environmental chemicals especially during developing period and natural environmental chemicals such as plant-derived odors have beneficial effects but some artificial chemicals have aversive effects on rat's behaviors.

1 Introduction

Brain functions have been shown to be influenced by various environmental chemicals. Natural environmental chemicals, such as plant-derived chemicals, modulate emotional state, nociception, autonomic functions and many other physiological functions [1]. On the other hand, certain artificial environmental chemicals impaired normal physiological functions, especially during development phases [2]. Recently endocrine disrupters have been shown to induce not only reproductive toxicity but also immune- and neurotoxicity. Industrial production of artificial environmental chemicals is increasing, while natural environmental chemicals are decreasing in the modern society. The brain functions seem to be

* Corresponding author.

A. Hanazawa et al. (Eds.): Brain-Inspired Info. Technology, SCI 266, pp. 79–84.
springerlink.com

under the influence of these changes in chemical environments, but detail is unknown. In this study, we investigated the effects of natural and artificial environmental chemicals on rodent's behaviors.

2 Materials and Methods

2.1 Animals

Six-weeks-old Wistar rats were used in this experiment. Rats housed on 12/12 LD cycle (light on at 8:00) with food/water available ad libitum. All experiments were performed under the Guidelines for Animal Experiments of the Graduate School of Life Science and Systems Engineering, Kyusyu Institute of technology and Low (No.105) and Notification (No.6) of the Government of Japan.

2.2 Exposure to Natural Environmental Chemicals

We used green odor and jasmine tea odor. Green odor, in which cis-3-hexanol and trans-2-hexenal are main substances, is released from green leafs. Jasmine tea is a typical flower tea that was originally produced in china and is now one of the most popular flower teas in Japan. Jasmine tea has been used in traditional Chinese medicine to activate the gastrointestinal function. Adult male rats (8-11weeks of age) were exposed to 0.03% green odors (mixture of cis-3-hexanol and trans-2-hexanal) for 5min or jasmine tea odor for 1min, then behavioral measurements were performed.

2.3 Exposure to Artificial Environmental Chemicals

We used 1-bromopropane (1-BP) and bisphenol-A (BPA). 1-BP is an ozone-depleting substance replacement. BPA, a typical endocrine disrupter, is used in epoxy resin and polycarbonate resin coating. Pregnant rats were exposed to 400ppm of 1-BP through the exposure chamber for 20 days (6 hours per day) until the day of birth (postnatal day 0, PND0). On the other hand, 0.1ppm BPA was exposed for 7days before birth. Dams were allowed to feed and drink ad libitum, until their pups were weaned on PND21. Pups were performed behavioral measurements after the age of six weeks. Behavioral measurements except for forced swimming test were performed in juvenile period (6-7 weeks). Forced swimming test was performed in adulthood period (12 weeks).

2.4 Open-Field Test

Rats were placed in the open field environment (80×60×40 cm) in a room and allowed to explore for 10min. The animals were monitored from above by a video camera connected to a computer running video tracking software (Ethovision v 1.96; Noldus Info.Tech.) to determine the time, distance moved, and rearing frequency.

2.5 Elevated Plus Maze Test

Rats were placed in the center of an elevated plus maze which elevated 50cm in a room and allowed to explore for 5min. The animals were monitored from above by a video camera connected to a computer running video tracking software (Ethovision v 1.96; Noldus Info.Tech.) to determine time spent in the open and closed arms, and the number of entries into the open and closed arms.

2.6 Forced Swimming Test

Rats were placed in a apparatus of water (24℃) to a depth of 35cm. The rats remained in the water for 15min and were then removed and allowed to dry in before returning to their home cage. The rats monitored from the above by a video camera, and data were stored on a videotape for later analysis. The tests were scored for duration of immobility.

3 Results

3.1 The Effect of Natural Environmental Chemicals

The green odor did not affect sensory-motor and emotional functions in open-field test (Fig. 1a) or elevated plus maze test (Fig. 1b) but locomotor activity and exploratory behavior in the open-field in rats were greatly increased after swimming for 15min (Fig. 1c-d). The jasmine tea odor increased distance moved and rearing frequency in the open field test (Fig. 2a-b) without affecting anxiety level evaluated by the elevated-plus maze test (Fig. 2c-d).

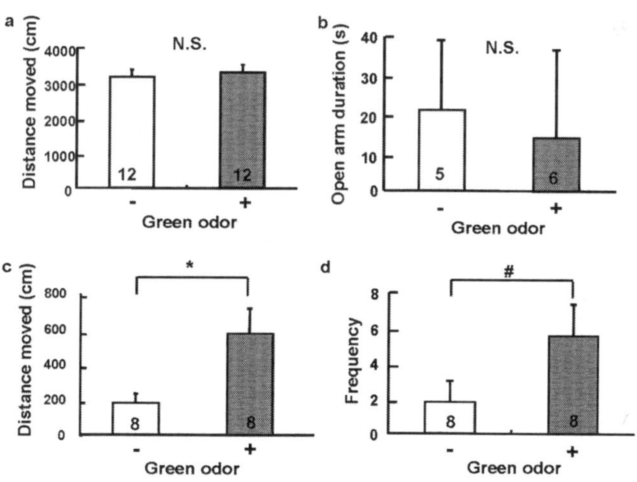

Fig. 1. Effects of green odor in behavioral measures

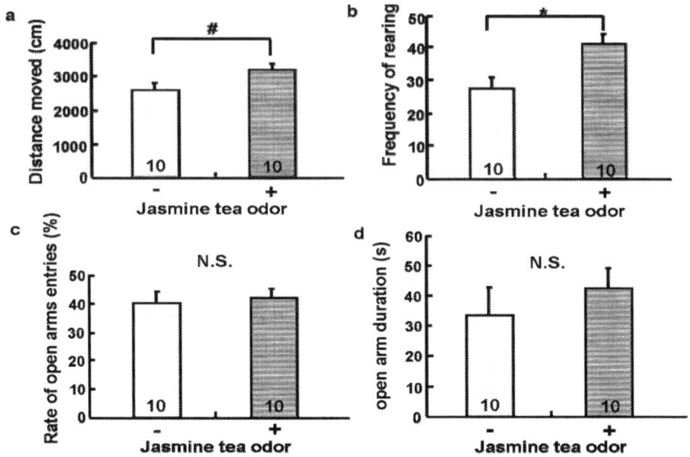

Fig. 2. Effects of jasmine tea odor in behavioral measures

The white and gray columns are expressed as mean ± SEM for the control group and green odor group data, respectively. *P<0.05, #P<0.1, Student's *t*-tests were used for between the control and green odor group. Five minutes exposure to green odor didn't affect a) distance moved (locomotor activity) in the open-field test and b) open arm duration (anxiety level) in the elevated plus maze test. But c) distance moved and d) frequency of entry the center area in the open-field test was increased after swimming for 15min.

The white and gray columns are expressed as mean ± SEM for the control group and jasmine tea odor group data, respectively. *P<0.05, #P<0.1, Student's *t*-tests were used for between the control and jasmine tea odor group. One minutes exposure to jasmine tea odor increased a) distance moved and b) frequency of rearing in the open field test. But c) Rate of open arm entries and d) open arm duration in the elevated plus maze test didn't affect.

3.2 The Effect of Artificial Environmental Chemicals

In the open-field test, females showed more frequency of rearing behavior than males in control group, while intrauterine exposure to 1-BP abolished this gender difference (Fig. 3a). In the elevated plus maze test which evaluated anxiety level, females showed more frequency of staying in closed arm than male in control group (Fig. 3b). On the other hand, the prenatal BPA exposure impaired the sexual differentiation of an exploratory behavior in the open field test (Fig. 4a), and enhanced depressive behavior in the forced swimming test (Fig. 4c) without affecting anxiety level evaluated by elevated plus maze test (Fig. 4b).

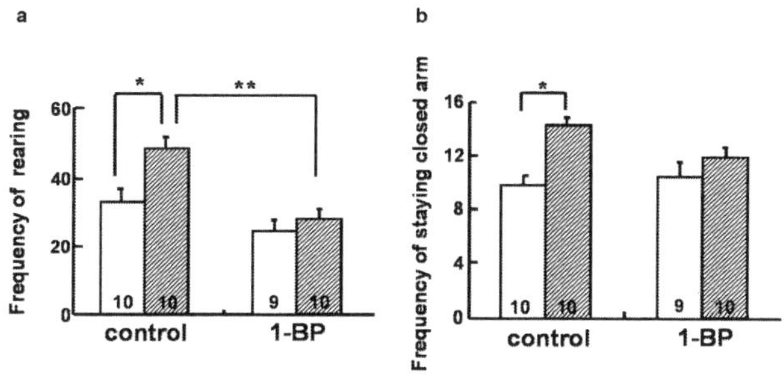

Fig. 3. Effects of 1-BP in behavioral measures

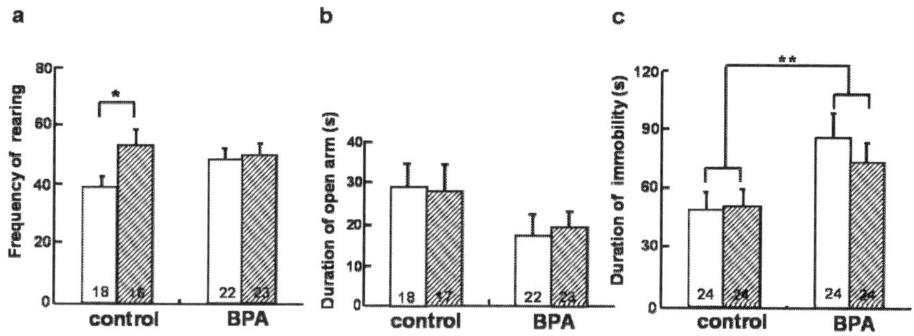

Fig. 4. Effect of BPA in behavioral measures

The white and shaded columns are expressed as mean ± SEM for the males and females data, respectively. *P<0.05, two-way ANOVA followed by Fisher's PLSD test between the male and female in same group. **P>0.05, two-way ANOVA followed by Fisher's PLSD test between the data of control and 1-BP group in same sex. Prenatal exposure to 400ppm 1-BP impaired sexual differentiation of a) frequency of rearing in the open-field test and b) frequency of staying in the closed arms.

The white and shaded columns are expressed as mean ± SEM for the males and females data, respectively. *P<0.05, two-way ANOVA followed by Fisher's PLSD test between male and female in same group. **P>0.05, two-way ANOVA followed by Fisher's PLSD test between the data of control and BPA group in same sex. Prenatal exposure to 0.1ppm BPA impaired the sex difference of a) frequency of rearing in the open-field test, but b) staying in the duration of open arms in the elevated plus maze test was not affected. BPA increased c) the duration of immobility in the forced swimming test.

4 Discussion

The results suggest that animal behaviors are highly sensitive to environmental chemicals especially during the development period. Natural environmental chemicals released from plants have beneficial effects. But some artificial environmental chemicals have adverse effects such as impairments of sexual differentiation of behaviors and facilitation of negative emotion. Since normal inputs of sensory information are essential to develop normal brain functions, chemical environments in the developing brain are also important for normal brain functions.

Acknowledgement

This research was supported by a COE program (center #J19) granted to Kyusyu Institute of Technology by MEXT of Japan and Grants-in-Aid for Scientific Research of the Ministry of Education, Japan (S.A.).

References

1. Aou, S., Mizuni, M., Matsunaga, Y., et al.: Chemical Senses 30, 262–263 (2005)
2. Fujimoto, T., Kubo, K., Aou, S., et al.: Brain Research 1068, 49–55 (2006)

Category Recognition in the Monkey Orbitofrontal Cortex

Takao Inoue[1,*], Balázs Lukáts[1], Tomohiko Fujimoto[1], Kotaro Moritake[1], Takeshi Hasegawa[1], Zoltán Karádi[2], and Shuji Aou[1]

[1] Department of Brain Science and Engineering, Graduate School of Life Science and Systems Engineering, Kyushu Institute of Technology, 2-4 Hibikino, Wakamatsu, Kitakyushu 808-0196, Japan
inoue-takao@edu.brain.kyutech.ac.jp
[2] Institute of Physiology and Neurophysiology Research Group of Hungarian Academy of Sciences, Pécs University, Medical School, Pécs H-7624, Hungary

Abstract. The orbitofrontal cortex (OBF) is known to play important roles in evaluation of reward value based on integration of multimodal sensory inputs. In this study, we investigated the complex cognitive functions in the OBF neurons from rhesus monkeys *(Macaca mulatta)*. Glass-coated Elgiloy electrodes and tungsten fiber multi-barreled glass microelectrodes have been used for extracellular recording during behaving the visual discrimination task. As a result, Neurons showed a wide variety of selectivity for these complex images. OBF neurons are involved in various levels of categorization and integrate various endogenous humoral signals and specific exogenous sensory cues.

1 Introduction

The orbitofrontal cortex (OBF) is known to play important roles in evaluation of reward value based on integration of multimodal sensory inputs [1]. Our previous studies demonstrated specific orbitofrontal cortical processes involved in adaptive regulation of feeding and emotional behaviors [6]. Neural mechanisms of category recognition of objects have been intensively studied in the inferotemporal cortex [7], [8]. However, those of biologically significant objects such as foods or heterosexual partners are poorly understood. To elucidate these complex cognitive functions in the orbitofrontal cortical neurons, electrophysiological studies were performed in rhesus monkeys *(Macaca mulatta)*.

2 Materials and Methods

2.1 Subjects and Data Acquisition

Macaque monkeys *(Macaca mulatta)* were used in this study. The experiment was performed in a dark and sound-attenuated room, with the monkey seated in a

* Corresponding author.

A. Hanazawa et al. (Eds.): Brain-Inspired Info. Technology, SCI 266, pp. 85–89.
springerlink.com

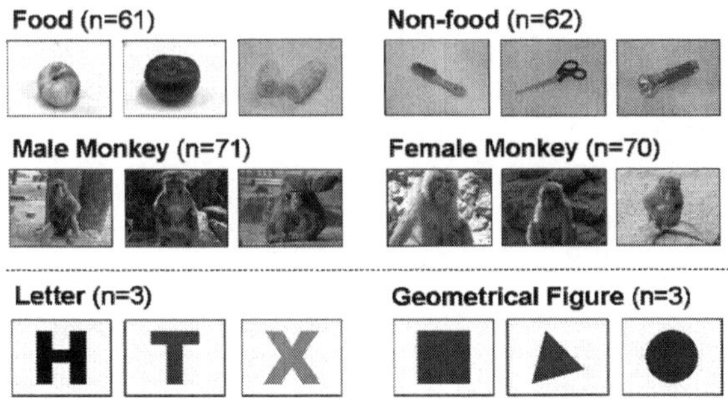

Fig. 1. Example of pictures used. The numbers in parentheses indicates the number of 2-dimensional images

primate chair. After the training had been completed, surgery was conducted. 4 anchor screws were implanted on the occipitoparietal region of the skull under pentobarbital anesthesia and aseptic conditions, and then a plastic plate for stereotaxic head fixation was mounted on the screws and fixed with dental cement.

Extracellular single neuronal activity was recorded by means of glass-coated Elgiloy electrodes or tungsten fiber multi-barreled glass microelectrodes in the OBF 1) in the course of visual discrimination tasks. Three paired sets of visual categories consisted of: (i) foods and non-foods; (ii) male and female monkeys; (iii) geometrical figures and letters were used as visual cues (Fig.1), 2) during microiontophoretic administration of various neuroactive substances, i.e. glucose, catecholamines and other neurochemicals (glutamate, GABA, ACh, etc.).

2.2 Behavioral Task and Data Analyses

Each trial of the task was started with a visual presentation of one of the following pictures stored in a computer. In food/non-food task, a picture either food or non-food objects was randomly presented. In male/female task, a picture from one of the sexes of Japanese monkeys was randomly presented. Geometrical figure/letter task were used as the control. Pre-determined target, for example category of male, was assigned on a lever, and the monkey has to press the lever when the "target" category, male in this case, was presented on a screen to get water reword. Tasks were composed of blocked-fashion, and in each block, a monkey had to discriminate the category of a presented picture. During learning of the target category, monkeys understood the pre-determined category within the first 10 to 20 trials.

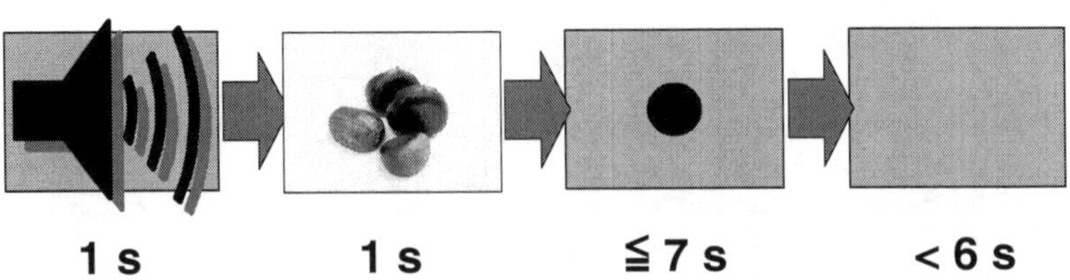

Fig. 2. Temporal sequence of visual discrimination task

In the data analysis, we examined neuronal activities during the visual stimulus presentation period (the period from 100 to 600 ms after the visual stimulus onset) on the discrimination task.

3 Results

Neurons showed a wide variety of selectivity for these complex images (Table 1). Although many neurons responded to all or some category in non-specific manner, about 5-10% of recorded neurons responded in a category-specific (Fig. 3) or task-specific manner, e.g. responded to only food category but not others or to food/non-food task but not other tasks (Table 1). In addition to these responses, expectation- and/or reward-related activities were also recorded. Among the category specific neurons, food selective neurons were the most predominant subgroup. OBF neurons were also shown to change in activity in response to various microiontophoretically applied chemicals (not shown). The majority of these 'chemosensory' cells exhibited differential sensitivity to glucose, monoamines, catecholamines, and other neurochemicals (glutamate, GABA, ACh, etc.).

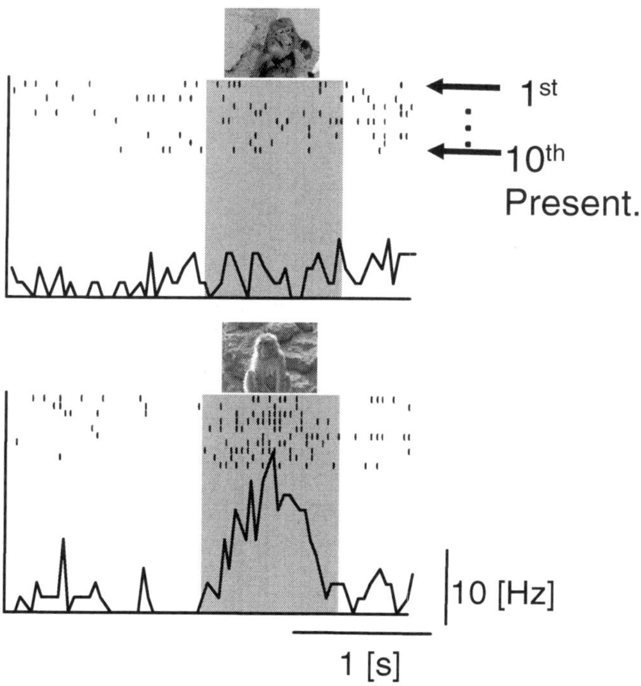

Fig. 3. Female category-selective neuron in the OBF. The vertical calibration represents 10 spikes/s, and the horizontal calibration represents 1.0 s. The neuronal data of the shaded period indicates presentation of stimulus picture. Upper: an example response to male picture. Lower: an example response to female picture Firing rate selectively increase during presentation of Female pictures.

Table 1. Stimulus selectivity of the OBF neurons

	Classification	Number of neurons
Category selective neurons	Food	20
	Non-food	6
	Food & Non-food	10
	Male	13
	Female	9
	Male & Female	9
	Food & monkey	4
	Figure & Letter Related	8
Other neurons	Cue*	9
	Reward Related*	35
	Sample Stimulus	22
	Target Category	11
	Non-target Category	5

*, some neurons showed category selective responses which also classified as category selective neurons.

4 Discussion

In this study, complex cognitive functions in the OBF are examined using extracellular recording and microiontophoretic application of neurochemicals. Our results provide evidence for the existence of special chemosensory neurons and hierarchical category neurons in the primate OBF. These neurons are involved in feeding and sexually motivated behavioral responses. These results are in connection with previous findings which provided evidence for the existence of feeding-associated neurons in the rodent OBF [3]. Therefore, OBF neurons underlie feeding and motivated functions preserved in a wide range of mammals. In general, categorization is the basis of conceptualization. It is interesting hypothesis that the monkeys conceptualize biologically significant objects such as food and gender. Santos et al [6] demonstrated that rhesus monkeys possess a domain-specific mechanism for categorizing the edible objects which seems to be the basic process in conceptualization of foods. Gender categorization is based on characteristic cues and global features which can be used for domain-specific mechanism [1, 2, 4]. In our preliminary study, we found category-selective OBF neurons for food and gender from a naïve untrained monkey. These findings propose that categorization of food and gender is acquired before training and/or innately.

5 Conclusion

This study demonstrated that neurons in the OBF are involved in various levels of categorization from individual category such as food, non-food, male and female to higher levels of categorization such as non-animal objects, sex. Moreover, OBF neurons integrate various endogenous humoral signals and specific exogenous sensory cues utilizing differential neurotransmitter mechanisms in the integrative-associative processes of the organization of adaptive behavior.

Acknowledgement

This work was supported by a COE program (center #J19) granted to Kyushu Institute of Technology by MEXT of Japan.

References

1. Eaton, G.G., Resko, J.A.: Ovarian hormones and sexual behavior in Macaca Nemestrina. J. Comp. Physiol. Psychol. 86, 919–925 (1974)
2. Koba, R., Izumi, A.: Sex categorization of conspecific pictures in Japanese monkeys (Macaca fuscata). Anim. Cogn. 9, 183–191 (2006)
3. Lukats, B., Egyed, R., Karadi, Z.: Single neuron activity changes to interleukin-1beta in the orbitofrontal cortex of the rat. Brain Res. 1038, 243–246 (2005)
4. Mizuno, M.: Visually-guided discrimination and preference of sexuality in female macaque monkeys. Fukuoka Igaku Zasshi 88, 105–116 (1997)
5. Rolls, E.T., Critchley, H.D., Treves, A.: The representation of olfactory information in the primate orbitofrontal cortex. J. Neurophysiol. 75, 1970–1981 (1996)
6. Santos, L.R., Hauser, M.D., Spelke, E.S.: Recognition and categorization of biologically significant objects by rhesus monkeys (Macaca mulatta): the domain of food. Cognition 82, 127–155 (2001)
7. Sigala, N.: Visual categorization and the inferior temporal cortex. Behav. Brain Res. 149, 1–7 (2004)
8. Vogels, R.: Categorization of complex visual images by rhesus monkeys. Part 2: single-cell study. Eur. J. Neurosci. 11, 1239–1255 (1999)
9. Yamamoto, T., Oomura, Y., Nishijo, H., et al.: Monkey orbitofrontal neuron activity during emotional and feeding behaviors. Brain Res. Bull 12, 441–443 (1984)

An Adaptive Controller System Using mnSOM (2nd Report: Implementation into an Autonomous Underwater Robot)

Yasunori Takemura[1,*], Makoto Ishitsuka[1], Shuhei Nishida[2], Kazuo Ishii[1], and Tetsuo Furukawa[1]

[1,*] Department of Brain Science and Engineering, Graduate School of Life Science and Systems Engineering, Kyushu Institute of Technology,
2-4 Hibikino, Wakamatsu, Kitakyushu 808-0196, Japan
takemura-yasunori@edu.brain.kyutech.ac.jp
ishitsuka-makoto@edu.brain.kyutech.ac.jp
{ishii,furukawa}@brain.kyutech.ac.jp

[2] Nagaoka University of Technology,
1603-1 Kamitomioka, Nagaoka, Niigata, 940-2188, Japan
nishida@pelican.nagaokaut.ac.jp

Abstract. Autonomous underwater vehicles (AUVs) have great advantages for activities in deep oceans [1], and are expected as the attractive tool for underwater development or investigation near future. However, AUVs have various problems which should be solved such as motion control, acquisition of sensors' information, behavioral decision, navigation without collision, self-localization and so on. Therefore, the AUVs should be autonomous and adaptive to their environment.

In this paper, a new self-organizing decision making system for AUVs using modular network Self-Organizing Map (mnSOM) [4] proposed by Tokunaga et al., is described. The proposed decision making system is developed using recurrent Neural Networks type mnSOM and the efficiency of the system is investigated through the simulations. And, we report that the decision making system is implemented into an autonomous underwater robot "Twin-Burger" (Fig. 1)[5].

1 Introduction

Autonomous underwater vehicles (AUVs) have great advantages for activities in deep oceans [1], and are expected as the attractive tool for underwater development or investigation near future. However, AUVs have various problems which should be solved such as motion control, acquisition of sensors' information, behavioral decision, navigation without collision, self-localization and so on. In order to realize the useful and practical robots which can work in the ocean, underwater vehicles should take their action by judging the changing condition from their own sensors and actuators, and are desirable to make their behaviors with limited efforts of the operators, because of the features caused by the working environment. Therefore, the AUVs should be autonomous and adaptive to their

* Corresponding author.

A. Hanazawa et al. (Eds.): Brain-Inspired Info. Technology, SCI 266, pp. 91–96.
springerlink.com

environment. We have been investigating applications of neural network technology into the AUVs focusing on the capability of neural networks (NNs) such as learning, nonlinear mapping, and proposed several controller and navigation methods using NNs and SOM. AUVs have nonlinear coupled dynamics in six degrees of freedom, and the changes of the equipments of robots have influence on the control system. In the previous adaptive control method in [2] and [3], the information of initial knowledge is getting less gradually during the process of adaptation. Therefore, a new method, which keeps the information of initial knowledge or previous environment and adapts to new environment, should be developed for improving the efficiency of the learning and reducing the learning cost with the use of the former environmental information that the robot had already learned.

In this paper, a new self-organizing decision making system for AUVs using modular network Self-Organizing Map (mnSOM) [4] proposed by Tokunaga et al., is described. The proposed decision making system is developed using recurrent NN type mnSOM and the efficiency of the system is investigated through the simulations. And, we report that the decision making system is implemented into an autonomous underwater robot "Twin-Burger" [5](Fig.1).

Fig. 1. Overview of Twin-Burger

2 Adaptive Controller Using mnSOM

The controller of the robot in a module consists of a Recurrent Neural Network (RNN). The adaptive controller is realized using RNN-mnSOMs; Forward Model Modules (FMMs) which express dynamics of the robot, and Controller Modules (CMs) which control the corresponding FMM. Figure 2 shows the architecture of control system using FMMs and CMs. The making processes of control system have following three steps. (a) Identification fo Forward Model Modules, (b) Adaptation of Controller Modules using the Forward Model Modules and (c) Implementation of the Control Module to Robot Control.

(a) *Identification of Forward Model Modules*
At first, FMMs are acquired several time series of motion data, which represents different dynamics of the robot are given to FMMs to represents one module a

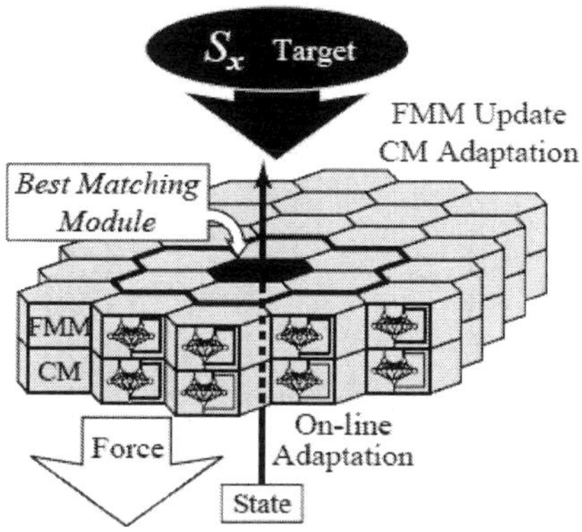

Fig. 2. Implementation of Control Module to Robot Control

dynamics property in advance. These time series data are fed into RNN-mnSOM, and FMMs are obtained.

(b) *Adaptation of Controller Modules using the Forward Model Modules*
Secondly, CMs are acquired using the fixed FMMs which are obtained by process (a). The target states variables are given to CMs and output data (control signals) calculated in CMs are given to all FMMs. The optimization of CMs is carried out by back-propagation method using the square error between target states and estimated states of FMMs regarding a FMM and a CM as one NN.

(c) *Implementation of the Control Module to Robot Control*
A Best Matching Module(BMM) among FMMs is decided by the observed robot State S_x (velocity, acceleration and control force). The CM which has the same index of the BMM in FMMs (black hexagon in Fig.2) is fed into target states, and the output force of the CM is given to the robot.

The adaptive controller using mnSOM is realized acoording to the processes (a)-(c).

3 Simulation of Controller

The simulations to examine the adaptability to unlearned data are carried out by comparing an adaptive controller in ref. [3] (hereafter, reference system) and this proposed controller is carried out. Figure 3 shows the transition of evaluation values. In these graphs, the horizontal axis means learning steps and vertical axis is evaluation value on log-scale. The upper figures are Forward Model Errors and the lowers are controller Errors. Solid lines are obtained from the proposed system and dot-lines are from the reference system.

On the reference system, the evaluation value of the forward model becomes big at the early stage of learning, then, it becomes smaller according to decrease of forward model error, the controller error becomes big. And, the controller can adapt to inputs at 20[sec] in the case of $M = 75$, and at 125[sec] in the case of $M = 120$.

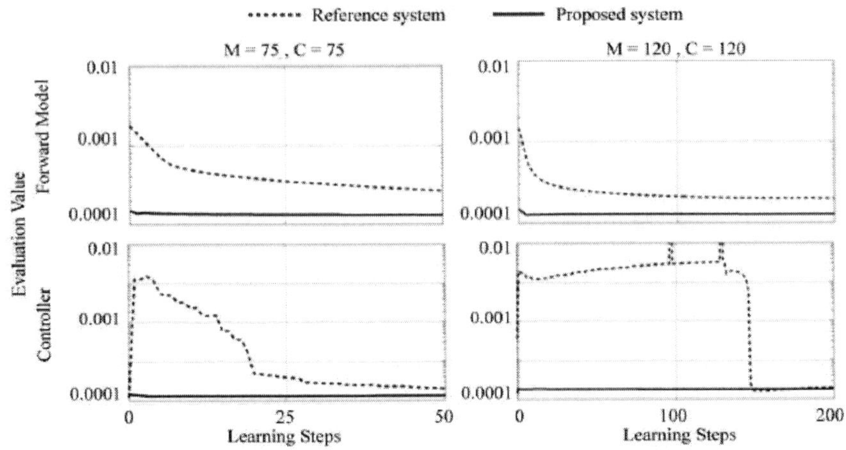

Fig. 3. Transition of Evaluate Values

On the proposed system, the forward model module which expresses given dynamic property exits, therefore, at the early stage of leaning, the proposed system needs few adaptations. The adaptability of proposed system is better than the reference system.

4 Experiment of FMMs Model Using AUV Data

Figure 4 shows the feature map of FMMs yaw directional controllers using mnSOM. Several time series of data are made from simulation by using eq.(1).

$$M\dot{r} + Cr|r| = F \tag{1}$$

Where, M is mass and added mass, C is coefficient of drag, r describes angular velocity of yaw direction and F is control force. Underwater robots motion equation are determined by eq.(1). Then, we simulated on limit cycle experiments using the learning data. Table 1 shows value of M and C in yaw direction on simulation data.

Table 1. Parameters of simulation value

M \ C	10	30	50	100
10	D_0	D_1	D_2	D_3
30	D_4	D_5	D_6	D_7
50	D_8	D_9	D_{10}	D_{11}
100	D_{12}	D_{13}	D_{14}	D_{15}

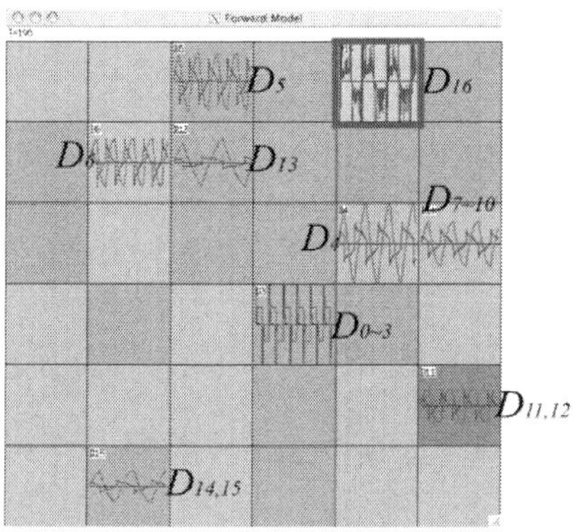

(a) Training times: 100

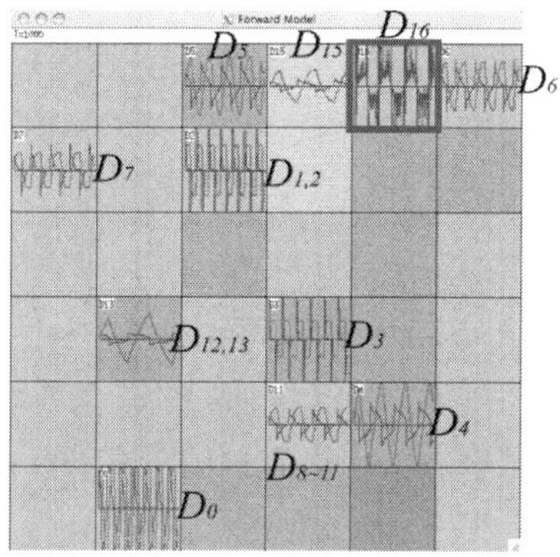

(b) Training times: 1000

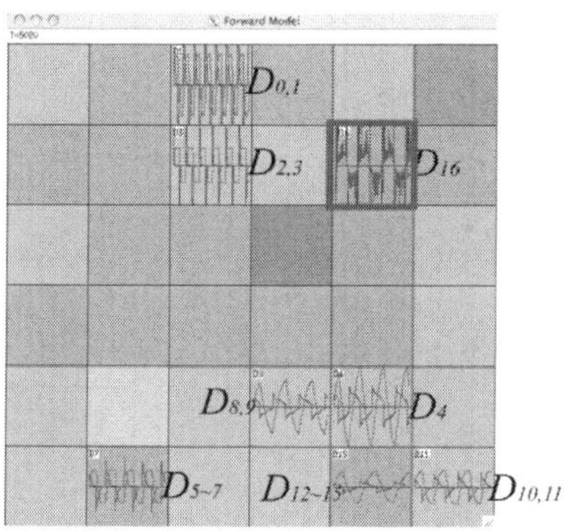

(c) Training times: 5000

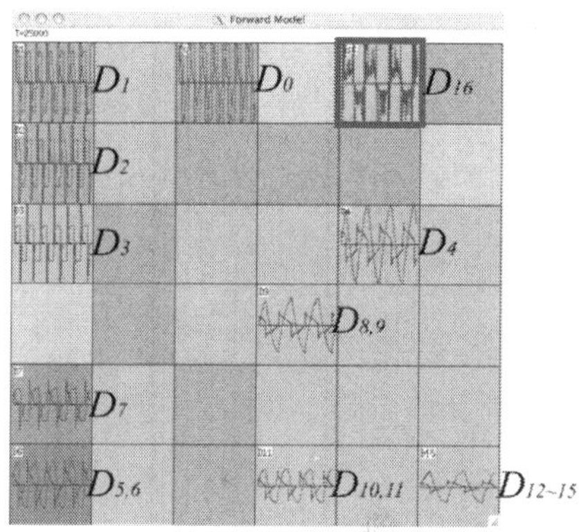

(d) Trainign times: 25000

Fig. 4. Feature map of FMMs mnSOM

Test data (no learning data) gets limit cycle experiment using "Twin-Burger". Considering Fig. 3, near parameter data (*M* and *C* parameter) located neighbor module in feature map. At that time, D_{16} class of experiment data located between D_0 and D_4. So that, D_{16} class parameters may be following number. *M* is between 10 and 30, and *C* is 10.

5 Conclusions

In this paper, we propose a new controller system using the mnSOM. This controller system is implemented into an autonomous under water robot "Twin-Burger". This controller system adapts quickly for unlearned states. The efficiency of the proposed system is investigated through the experiments using AUV. This system becomes tuned into various environments and dynamics.

Acknowledgement

This work was supported by the 21st Century Center of Excellence Program, "World of Brain Computing Interwoven out of Animals and Robots (PI: T.Yamakawa)," granted in 2003 to the Department of Brain Science and Engineering (Graduate School of Life Science and Systems Engineering), Kyushu Institute of Technology, by Japan Ministry of Education, Culture, Sports, Science and Technology.

References

1. Ura, T.: Free Swimming Vehicle PTEROA for Deep Sea Survey. In: Proc. of ROV 1989, pp. 263–268 (1989)
2. Kohonen, T.: Self-organized formation of topologically correct feature maps. Biological Cybernetics 43, 59–69 (1982)
3. Ishii, K., Ura, T.: An adaptive neural-net controller system for an underwater vehicle. Journal of IFAC Control Engineering Practice 8, 177–184 (2000)
4. Tokunaga, K., Furukawa, T., Yasui, S.: Modular Network SOM: Extension of SOM to the realm of function space. In: WSOM 2003, pp. 173–178 (2003)
5. Ishitsuka, M., Ishii, K.: Development of an underwater manipulator mounted for an AUV. In: CD-ROM Proc. of Oceans 2005, 6 pages (2005)

An Environment Recognition System Using SOM

Atushi Kanda, Masanori Sato, and Kazuo Ishii

Department of Brain Science and Engineering, Graduate School of Life Science and Systems Engineering, Kyushu Institute of Technology,
2-4 Hibikino, Wakamatsu, Kitakyushu 808-0196, Japan
kanda-atushi@edu.brain.kyutech.ac.jp

Abstract. Recently, various mechanisms have been developed combining linkage mechanisms and wheels. Especially passive linkage mechanisms and small wheel type robots are proposed, because a wheel type mobile mechanism has difficulty on the rough terrain movement. In our research, a 6-wheeled mobile robot employing the linkage mechanism has been developed to enhance maneuverability, and achieved climbing over a 0.20[m] height of bump. In this paper, we propose a switching controller system for wheel type mobile robot on rough terrain environment. This system consists of two sub systems. One is the environment recognition system using self-organizing map. The other one is the adjusted controllers using neural networks.

1 Introduction

The mobile mechanism using wheels is one of the most popular mechanisms for mobile robots because the wheel type mobile mechanism's energy efficiency is high, the mechanism is simple and the control system is well investigated. However, wheel type mobile robots have difficulties in rough terrain movements. In other words, the height of bump where a robot is able to climb over without inertia force for the normal wheel type mobile system is generally 1/3 or less of its wheel diameter. As a solution for the difficulties, link mechanisms with small wheels to realize high mobility have been developed. For example, NASA/JPL developed Rocker-Bogie mechanism and installed into Sojourner [1], Kuroda et al. developed PEGASUS mechanism and installed into Micro5 [2], EPFL developed linkage mechanisms and installed into Shrimp [3]. From the point of software, the robots need adaptive controllers in the various environment using sensors for environment recognition, laser-range finder and stereo camera, etc.

In the previous research, we developed a 6-wheeled mobile robot employing the linkage mechanisms, "Zaurus", to evaluate maneuverability of wheel type robots. Figure 1 shows overview of Zaurus and Fig. 2 shows degrees of freedom of Zaurus. The performance of controller for rough terrain using a neural network and PID controller is compared [4]. The neural network controller showed almost same performance with the well-tuned PID controller.

In this paper, we proposed environment recognition system that selects the adjusted controller for each environment based sensory data. The input data are

A. Hanazawa et al. (Eds.): Brain-Inspired Info. Technology, SCI 266, pp. 97–102.
springerlink.com

Fig. 1. Overview of Zaurus

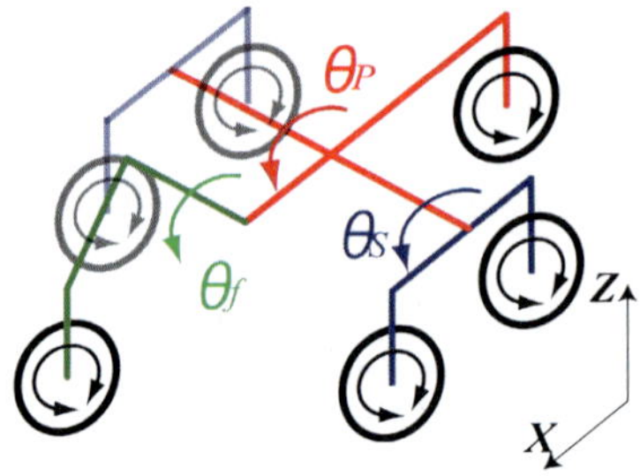

Fig. 2. DOF of Zaurus

the linkage angles and fed to Self-Organizing Map(SOM) which is the basic information processing element of the environment recognition system.

2 Environment Recognition System

The proposed system needs the function to classify environment using the linkage data and estimate current terrain. Figure 3 shows concept of environment recognition system. This system recognizes various environments using data of passive link joints.

2.1 Basic Environment Data

The input data to the system includes dynamics of the robot. In this paper, we defined environment recognition as "Classifying environment of bump of height using joint angle data, making basic environment data, and identifying the traveling environment".

Experiments to make basic data set several time series of climbing and down of bumps, front angle(θ_f), side link angle(θ_s), attitude angle of pitch(θ_p) are used as teaching data. Basic environments are one-step series whose heights are 0.0[m], 0.06[m], 0.12[m] and 0.18[m], and a PID controller is employed for data sampling. Next, we describe the process of making teaching data. Basic state variables consist of three angles mentioned about and their angular velocities.

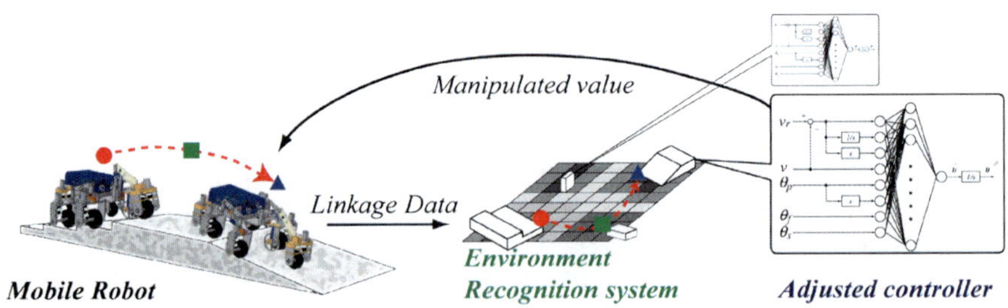

Fig. 3. The concept of environment recognition

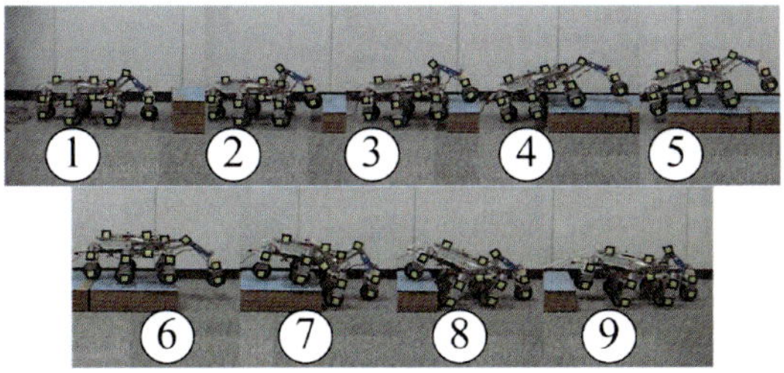

Fig. 4. Label of attitude of robot

The sampling cycles is 0.25[sec] and state various during 1.0[sec], that is 4 set of state become an input data.

Figure 4 shows the attitude of robot climbing. No.1 show driving on surface, No.2 through No.5 show climbing up, No.6 through No.9 show climbing down.

2.2 Principle Component Analysis: PCA

The principle component analysis (PCA) is one of the feature extraction algorithms for less deficit data. The advantages of PCA are the possibility of contraction of data and a feature that the extracted data are able to be evaluated independently because extracted data are independent.

Figure 5 and 6 shows the result of analysis using PCA, bump of height are 0.06, 0.12, and 0.18[m] respectively. Horizontal axis shows the first principle component and vertical axis shows the second principle component. In Fig. 6 the climbing up and down data does not show the same trend. Therefore, the PCA is difficulties to use as the environment recognition system.

2.3 k-Means Method

Regarding to the multiple classification analysis, k-means method is one of the most popular nonhierarchical analysis.

The number of cluster, *k*, is assumed, and each input data is classified into the similarity cluster. Then, the new center of the cluster is calculated and all data as re-classified into new clusters. Finally, the procedure described above is repeated

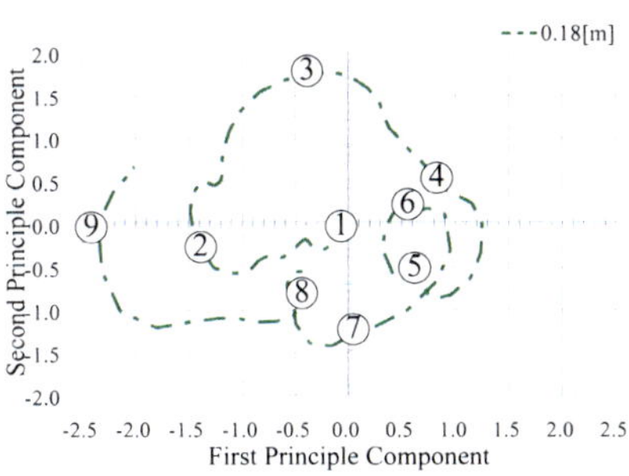

Fig. 5. The result of PCA: height of bump 0.18[m]

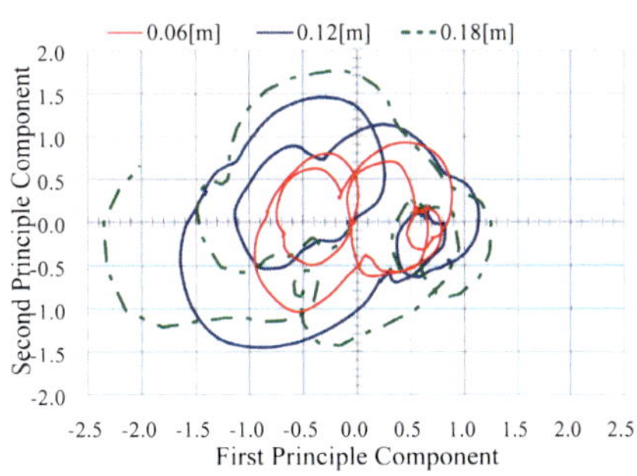

Fig. 6. The result of PCA : All data

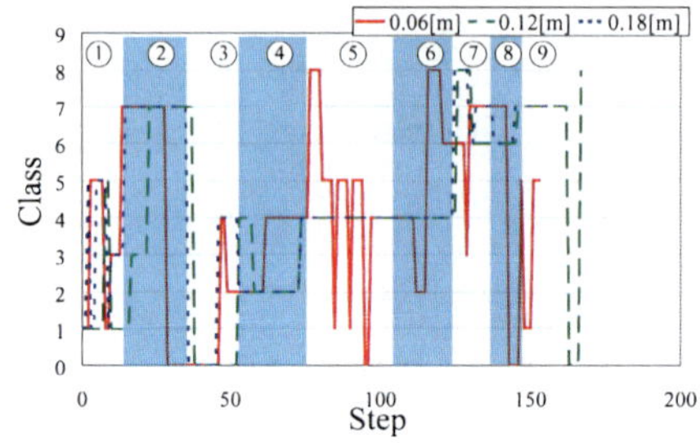

Fig. 7. The result of k-means : cluster number is 9

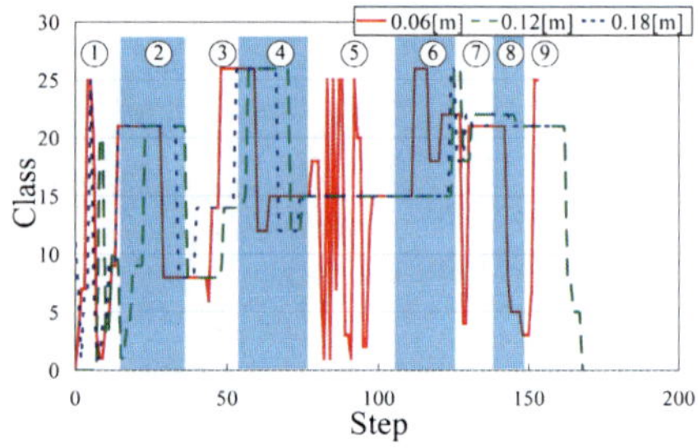

Fig. 8. The result of k-means : cluster number is 27

until the error between all points and the centers of clustered data become sufficiently small.

Figure 7 and 8 shows the results of analysis using k-means. The number of cluster is 9 and 27 repeatedly. Horizontal axis shows the sampling step and vertical axis shows the cluster number. The cluster number 9 is assumed that robot has nine situations. The cluster number 27 is assumed that nine situations and each height of bumps.

As shown in Fig. 7, the data classify as the attitude of robot climbing up and down. However, the height of bumps is not clustered because the cluster number 9 and cluster number 27 are almost same result.

2.4 Self-Organizing Map (SOM)

Self-Organizing Map proposed by T.Kohonen is one of well-known methods for classifying data with preserving topological feature map. The SOM is trained using unsupervised learning to produce low dimensional representation of training data.

The neighborhood functions (ϕ_i^k) of k-th unit to i-th input data is described employing neighborhood radius (σ) is eqs.(1)(2). The leaning coefficient (ψ_i^k) is normalized using eq.(3). k^* is the indication number of the winner unit.

$$\sigma = \sigma_{\min} + (\sigma_{\max} - \sigma_{\min})\exp(-t/\tau) \tag{1}$$

$$\phi_i^k = \exp\left(\left(-d(k,k^*)\right)/2\sigma^2\right) \tag{2}$$

$$\psi_i^k = \phi_i^k / \sum_i \phi_i^k \tag{3}$$

Figure 9 and 10 shows the obtained feature map and color of each unit means distance vector between the reference neighborhood units. The white color means far and the black color means that we can see that similar units locate closely. The specifications of map, map size is 30 x 30, learning time (t) is 1000, neighbor radiuses $\sigma_{\max} = 45$, $\sigma_{\min} = 1$, time constant (τ) is 50. The finish learning time is about 500.

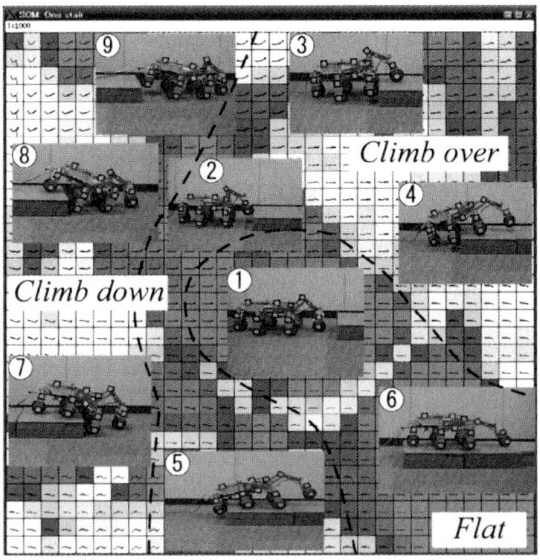

Fig. 9. The result of SOM: Classification about attitude of robot

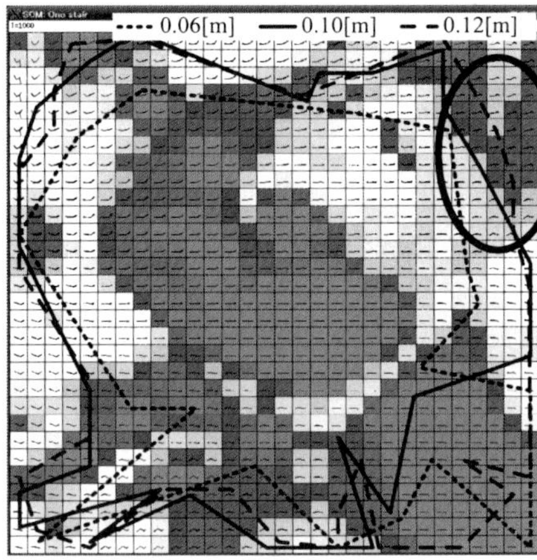

Fig. 10. The result of SOM: Evaluation of interpolation function

As shown in Fig.9, the map is classified based on the attitude of robot. The movement of climbing up and down is arranged in opposite side. Traveling on the flat floor is assigned to the center of the map. The feature map by SOM also includes the dynamics of the robot.

As shown in Fig.10, we evaluate interpolation function using unlearned experimental result (0.10[m] height). The trajectory of 0.10[m] height is similar to those of 0.06[m] and 0.12[m], and the unlearned data is assigned in the middle of those lines. The black circle signs the point. We can see that the map can estimate the attitude of robot and height of bump. The areas where the trajectories overlap are the conditions of where the differences are not clear.

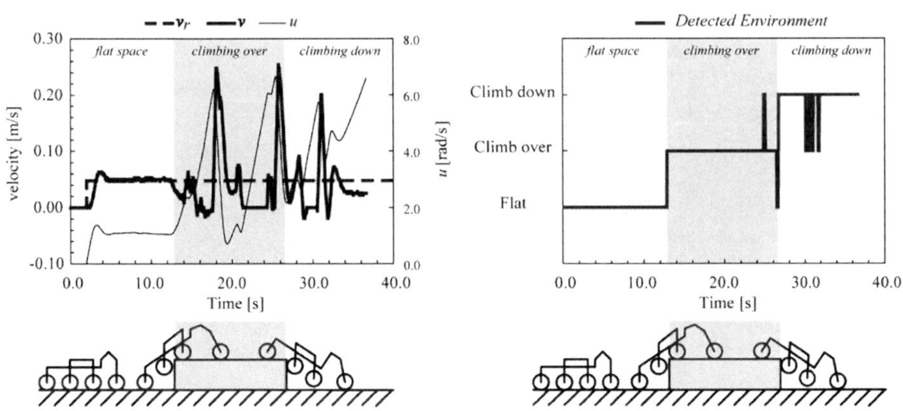

Fig. 11. The experimental result of environment recognition system

3 Online Recognition

The developed environment recognition system is introduced into Zaurus to evaluate the feature map classification online. Figure 11 the velocity of the robot and which environment the system categorized the current condition. The conditions are classified into 3 states, climb up and down and flat. The velocity of the robot is controlled properly and the environment recognition is working online.

4 Conclusion

In this research, we proposed the environment recognition system by classifying the angle data. The system consists of the two components, one is the environment recognition system using SOM, and the other is the adjusted controller using neural network. The basic environment map can classify attitude and height of bumps. The system is able to recognize the simple traveled terrain while traveling.

Acknowledgement

This work was partly supported by a grant of Knowledge Cluster Initiative implemented by Ministry of Education, Culture, Sports, Science and Technology (MEXT).

References

1. Volpe, R., et al.: The Rocky 7 Mars Rover Prototype. In: Proc. of IROS (1996)
2. Kuroda, Y., et al.: Mobility Performance Evaluation of Planetary rover with Similarity Model Experiment. In: Proc. of ICRA (2004)
3. Estier, T., et al.: An Innovative Space Rover with Extended Climbing Abilities. In: Proc. of Space a Robotics 2000 (2000)
4. Sato, M., Ishii, K.: A neural network based controller for a wheel type mobile robot. In: Brain-Inspired IT II 2006, pp. 261–264 (2006)

A Novel Embedded Computer Based Autonomous Robot for Sewer Pipe Inspection

Ahmad Ghaderi, Amir A.F. Nassiraei, and Kazuo Ishii

Department of Brain Science and Engineering, Graduate School of Life Science and Systems Engineering, Kyushu Institute of Technology,
2-4 Hibikino, Wakamatsu, Kitakyushu 808-0196, Japan
ghaderi@brain.kyutech.ac.jp

Abstract. In this paper, in order to realize an effective inspection system, an embedded computer based autonomous pipe inspection robot is presented. This robot is designed as a fully autonomous, un-tethered robot, which fits to the pipes within a diameter range of 200- 300 millimeters. The practical field tests show that the robot performs well and can do pipe inspection properly and the robot moves gracefully not only at straight paths but also at junctions and curves.

1 Introduction

Underground infrastructure is a major section of urban infrastructures and major cities invest considerable amounts of budget in their underground infrastructure including water and sewage pipes. Sewer systems are prone to damage due to aging, excessive traffic, geological change, earthquakes and chemically induced erosion. Due to leaks in the pipes caused from these damages, the groundwater is increasingly contaminated. Furthermore, sewage may wash away soil, possibly eroding the buildings foundations or the streets underground and pavements [1]-[2]. Thus, in order to ensure an optimal functioning sewer system, extensive inspection is necessary.

The rehabilitation of underground pipes is costly and slow in inspection process, because the size of most pipes eliminates the possibility of direct human inspection1 and requires employment of sophisticated equipment.

In current conventional pipes inspection methods, the inspection of sewer pipes is carried out using a tele-operated robot which has an on-board video camera system and is remotely controlled, via a tether cable, by a human operator. (Fig. 1) All equipment necessary to supply and control the robot are arranged inside of a car, out of the manhole which causes a heavy and stiff cable for the robot.

Because these robots are tethered to the control unit in the car, they are driven only from one manhole to the next, and then robots are driven backward through the pipe to they entry point, then they are retrieved, and the process is repeated again at the next manhole. Also these robots have a poor mobility function to pass any kind of pipe-bends such as curves and junctions. These two main reasons, “heavy and stiff cable” and “poor mobility function”, makes the inspection process difficult and costly [3].

A. Hanazawa et al. (Eds.): Brain-Inspired Info. Technology, SCI 266, pp. 103–108.
springerlink.com

To realize an autonomous pipe inspection system, mobile robot should carry all required resources on-board. Navigation should be performed completely by control programs running on on-board computing equipment. Status messages may be communicated to a human inspector over a radio link. Assessment of the pipe condition may be performed partially on-board, or offline after retrieval of the recorded sensory data.

Although a few researches have been done in development of a fully autonomous mobile robot [4]-[5], these robots are not yet marketable system today. Because the degree of development of complete autonomous sewer robots presently does not warrant using them safely and robustly in sewers. Most of the current robots have a complex moving mechanism and multi-sensor equipment for navigation and motion control. These complexities in mechanism and data processing make not easy to realize reliable commercial products especially for small range of the pipes up to 300 millimeter in diameter.

In this paper, in order to realize an effective inspection system, an embedded computer based autonomous pipe inspection robot is presented. In this robot, inspection efficiency is achieved by reducing the time and manpower in the inspection process. The robot presented in this paper is the prototype of a passive-active intelligent, fully autonomous, un-tethered robot which utilizes an embedded computer. Embedded computer has been used because it has many advantages likes robust construction, low power consumption, small size, compact form and reliability. In addition to realizing a fully autonomous inspection robot, we developed a small and intelligent 2D laser scanner for detection of the navigational landmarks such as manholes and inlets as stand-alone system and fusion with a fish eye camera to assist the pipe state and fault detection.

The practical field tests show that the robot performs well and can do pipe inspection properly and the robot moves gracefully not only at straight paths but also at junctions and curves.

2 Robot Architecture

This robot is designed as a fully autonomous, un-tethered robot, which fits to the pipes within a diameter range of 200- 300 millimeters. Robot autonomy and its kinematics abilities extend its potential mission range enormously, comparing to conventional inspection equipments that usually are limited by the cable and poor kinematics. The proposed robot which is shown in Fig.2 carries all the necessary resources on-board. Standard lithium polymer batteries provide the power for its 4 motors, the sensors, the light and electronics, including an embedded computer system and a fiber optic communication module.

Because Linux offers great advantages like as faster design and increased productivity and reliability, we used it as operating system. The controller is a single board computer that consists of a VIA Eden 660 MHz processor, 128 MB memory, 4 RS- 232 serial ports, 4 USB 1.1 ports and PC/104 ISA expansion BUS. It also has a data acquisition set with 16 analog inputs, 4 analog outputs, 24 programmable digital I/O and 2 programmable counter/ timer.

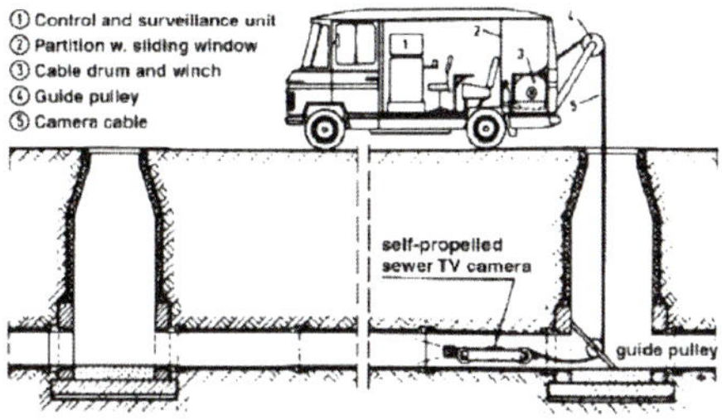

Fig. 1. Current conventional pipes inspection system

Fig. 2. Proposed inspection robot

Four high speed permanent magnet DC motors, which are controlled by four motor controllers, drive the robot. In this system motor shaft encoder data are used to calculate actual speed. The data are compared to reference speed after which the system makes proper pulse wide modulation (PWM) signals to supply motors.

In the communication between each motor driver and the embedded computer a RS232 interface is used. And necessary software is prepared to make a connection between robot and motors. In this software the connection properties like baud rate, type of parity and stop bits are defined.

The mechanical parts of robot like wheels, body, attachment, camera base, and suspension systems are designed to prepare a reliable structure of robot with good dynamic features.

This robot has infra red sensors that are derived using analog inputs of the embedded computer. Although the robot can be used autonomously, it is possible to control it manually via a connection to a host computer.

3 Robot Platform

With regard to the current sewer pipe inspection technology, all commercial robots are capable to move in straight pipes but not any kind of pipe-bends. In addition, current sewer inspection robots are not able to pass different size of pipes.

Although there have been a number of efforts and researches to develop a robot with capability of passing the pipe bends, most of these robots have a complex mechanism and several sensors for detecting the pipe bends and pass them. To avoid of these complexities in mechanism and data processing which makes not easy to realize reliable commercial products, in this paper, a passive active intelligent, compact and novel moving mechanism is employed.

"nSIR Mechanism" is a special patented mechanism [6] that can move into the straight pipes and passes different kinds of pipe bends without need to any intelligence of the controller or sensor reading. This passive-active mechanism itself can move into the pipe and passes wide variety of the pipe bends even without controller for the wheels actuators. In addition, this moving mechanism has capability to pass the different size of pipes in diameter even from a bigger diameter pipe to smaller diameter and also can pass a small obstacle and go down a small step. The design is based on the concept "passive adaptation of robot wheels to the bends in the pipe". This is accomplished by proper wheels orientation and passive damping of springs.

4 Robot Power System

The robot power system includes a Li-Polymer battery (25.9 [V]) and a 60 Watt multi output switching power supply and protection circuit. The necessary voltage for the camera, laser scanner, light, infrared sensors, underground communication, fiber optic communication, and the embedded computer power supply are converted from 25.9 [V] to 12.0 [V], 24.0 [V], -12.0 [V] and 5.0 [V]. The robot electrical storage unit must be sized by an energy storage and power requirement. Therefore the robot battery performance characteristics (Energy/weight Energy/size and Power/weight) should be enough to store sufficient energy and to provide the necessary peak power. After calculation of necessary power and energy for our robot we chose Lithium ion polymer batteries, because of its high power density (2220 W/kg) and high energy density (185 Wh/kg) [7].

This battery should not be discharged to a level below 3V per cell under load, because deep discharge below 3V per cell can deteriorate battery performance [8]. Therefore To avoid of battery damage, a protection circuit against deep battery discharge was designed and was employed in robot.

5 Fault Detection

This robot is capable to detect the candidate faulty images (called feature-images), by using a fast and simple edge detection program Feature-images may include three different types of image includes landmarks (manholes, pipe joints and inlets), faulty images (Fig. 4-left), non-faulty images. When robot detects a feature-image, an autonomous control program will start to decrease the robot speed for saving enough images, up to 5 frames, from feature-image area.

At this time, the laser scanner that has been kept switch off during the normal pipe condition will start to extract the possibility of happened feature fault as a land mark. In case of detection of fault as a land mark, the type of landmark, only as a code, will be saved via CAN interface into the hard disk. These start and stop process of saving data from camera and switch on/off the laser scanner perform a longer inspection process where we consider the capacity of hard disk and energy consumption. In addition, the time, robot speed and encoder data are saved to the hard disk during whole inspection process.

After the robot is retrieved, all data saved in the robot hard disk will transmit via the LAN interface to the other PC. In this PC, three different programs are applied to extract the real faulty images from the retrieval feature-images and calculate the position of happened fault. First, the feature-images are divided to the landmark images and the faulty-non faulty images by applying the laser scanner data and the time to the images. In the next step, special fault detection

Nine types of fault
crack
water infiltration
root invasion
pipe break
pipe gap
packing gap
mortar
an alien substance
landslide

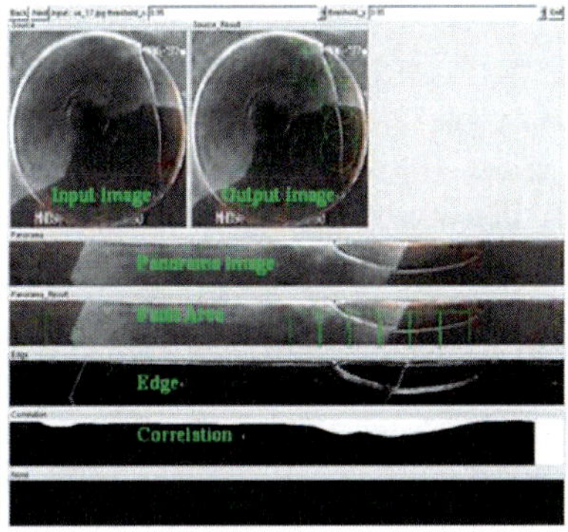

Fig. 3. Types of fault, possibly happening in the sewer pipe (left side). An example of result of special fault detection patented algorithm applied on a faulty image including a crack (right side).

patented algorithm extracts up to 60-70% of the faulty images from the faulty-non faulty images (Fig. 4) [9].

Last, the rough position of the faults will be calculated by using the encoder data, robot speed and time. This program has possibility to accept the map data of the sewer network (position of manhole, inlets and pipe joints) and fusion with laser scanner data to decrease the error of wheels slip and more accurate calculation of the happened fault position. After of all, remain faulty images including the necessary specification will save as a result of inspection process.

6 Conclusions

In this paper we proposed a fast and robust sewer inspection method by using a passive-active intelligent, fully autonomous, un-tethered robot that utilizes an embedded computer, which fits pipes within a diameter range of 200-300 millimeters. This robot can be driven by a control unit located in out of the manhole either via an optical underground wireless communication module, or via a life-optic-untethered cable.

The practical field tests show that the robot performs well and can do pipe inspection properly and the robot moves gracefully not only at straight paths but also at junctions and curves.

Acknowledgement

This research was supported by the FAIS-Robotics Research Institute, funded by the Organization for Small & Medium Enterprises and Regional Innovation, Japan (SMRJ). We thankfully acknowledge the cooperation with our partner: Yaskawa Electric Co., Kyushu Keisokuki co., Nihon Tecmo Ltd., FAIS and SMRJ.

References

1. Nassiraei, A.A., Kawamura, Y., Ahrary, A., et al.: Concept and Design of A Fully Autonomous Sewer Pipe Inspection Mobile Robot KANTARO. In: 2007 IEEE International Conference on Robotics and Automation (ICRA 2007), Roma, Italy, April 10-14 (2007)
2. Nassiraei, A.A., Kawamura, Y., Ahrary, A., et al.: A New Approach to the Sewer Pipe Inspection: Fully Autonomous Mobile Robot KANTARO. In: 32nd Annual IEEE Industrial Electronics Conference (IECON 2006), November 7-10 (2006)
3. Material provided by Steinbeis Japan Inc., Kitakyushu foundation forthe Advancement on Industry, Science and Technology (March 2002)
4. Kirchner, F., Hertzberg, J.: A prototype Study of an AutonomousRobot Platform for Sewerage System Maintenance. Autonomous Robots 4, 319–331 (1997)
5. Rome, E., Hertzberg, J., Kirchner, F., lichet, U., streich, H., Christaller, T.: Towards Autonomous Sewer robots: The MAKRO project. J. Urban Wate 1, 57–70 (1999)
6. Mechanism of movement inside of the pipes and mechanism of pipeinspection, Inventor: Amir A. F. Nassiraei, Japan Patent No. 2005-17384 (January 2005)
7. Ehsani, M., Gao, Y., Miller, J.M.: Hybrid Electric Vehicles: Architecture and Motor Drives. Proceeding of the IEEE, Electric, Hybrid & Fuel Cell Vehicles 95(4), 719–728 (2007)
8. `http://www.thunderpower-batteries.com/Li-PolyBatteries.htm` (Cited 5 December 2007)
9. A device and method for fault detection in sewer pipe system, Inventor: Alireza Ahrary, Japan Patent No. 2005-209677 (July 2005)

The Angular Threshold for Frog Collision Avoidance Behavior Changes Depending on Not Only the Stimulus Location But Also the Behavioral Strategy

Hideki Nakagawa[1,*] and Ryota Nishioka[2]

Department of Brain Science and Engineering, Graduate School of Life Science and Systems Engineering, Kyushu Institute of Technology, 2-4 Hibikino, Wakamatsu, Kitakyushu 808-0196, Japan
naka@brain.kyutech.ac.jp,
nishioka-ryoto@edu.brain.kyutech.ac.jp

Abstract. In our previous study, we showed that the frog, *Rana catesbeiana,* produces collision avoidance behavior when the retinal size of an approaching object reaches about 20°. In this study, we examined whether the threshold value is unchangeable under any conditions. The avoidance behavior of the frog to looming stimulus moving toward the animal on a collision course was examined by using computer graphics and frame by frame analysis. The distance between an animal and a display on which the looming stimulus was presented was changed from 20 cm to be 40cm. The mean angular thresholds are compared among three different areas on the screen of a display. Area B at a distance of 20 cm and area C at 40 cm subtended the same visual angle on the center of each display. Area A at a distance of 20 cm corresponds to outside the visual angle. Statistical analysis showed that the mean angular threshold in areas A and B were not significantly different ($p>0.05$). The mean angular thresholds in areas B and C were not significantly different, either ($p>0.05$). However, the analysis showed that the mean angular threshold in area C was significantly smaller than that in area A ($p<0.05$). These results suggest that the difference in angular threshold depending on the location of a display is caused by both visual field specificity and integration of size and depth information of an approaching object. We also examined the relationship between the angular threshold of visual stimulus and the direction of evoked escape jumps which is crucial for successful survival. The analysis showed that the angular threshold for a backward jump is significantly smaller than that for a forward jump. These results show that the angular threshold size can be changed depending on both stimulus condition and the behavioral strategy or prospective behavior performed by the animal.

[*] Corresponding author.

A. Hanazawa et al. (Eds.): Brain-Inspired Info. Technology, SCI 266, pp. 109–113.
springerlink.com

1 Introduction

Two different strategies for collision avoidance have been established in many animal species so far. The first strategy is to use time-to-collision as in the case of landing of the fly and pigeon, plunge dive in the gannet, driving in human. The second is to use an angular threshold of the retinal image as in the case of avoidance behavior of the fiddler crab, the chick and the locust. The crab and the chick display avoidance behavior when the retinal image of an approaching object reaches a visual angle of about 30° [5]. The threshold size for avoidance behavior of the locust is about 10° [4]. In our previous study, we showed that the frog produces collision avoidance behavior when the retinal size of an approaching object presented by using computer display reaches about 20° [6]. Are these angular threshold values invariable under any conditions?

In our preliminary experiment, we found that when the location of a display was changed from 20 to 40 cm, the angular threshold in the latter case (mean ± SD, 13.9 ± 6.6^{o}, n=36) was significantly smaller than that in the former case (mean ± SD, 17.5 ± 8.6^{o}, n=56) ($p<0.05$) [3]. There are two possible explanations to account for the difference in angular threshold due to the location of a display. The region where the looming stimulus can be presented changes depending on the location of a display. Therefore, the first is that the angular threshold shows visual field-specificity (Hypothesis 1). The second is that the frog processes not only retinal size but also depth information of an approaching object in the collision avoidance behavior (Hypothesis 2). To test these hypotheses, we compared the angular thresholds in different area on a display located 20 cm or 40 cm above a freely moving frog and examined whether the difference in angular threshold was caused by difference of stimulus location or difference of display height. Frog produces escape behavior in various directions in response to the looming stimulus above the animal. The direction of escape behavior is crucial for successful survival. Here, we examined the relationship between the escape direction and the location of visual stimuli. We also examined the relationship between angular threshold size and the stimulus location or direction of escape jump.

2 Materials and Methods

All experimental procedures were in accordance with Kyushu Institute of Technology Animal Care Use Regulations. The frog, *Rana catesbeiana*, was placed on a transparent experimental stage. Looming stimuli were presented in the dorsal visual field of the animal with a display placed 20cm or 40cm above the stage. They simulated a retinal image of a black square (35 x 35 cm) approaching at a velocity of 2m/s through a path of 6m. Evoked escape behavior was recorded by a substage CCD camera and a video cassette recorder at 30 frames per second. Each avoidance behavior was fed into a computer through an NTSC video capture card (SIM-PCI, DETECT) and saved as successive still video frames. To measure animal positions, the coordinates of the snout and the vent of the frog were obtained. The threshold of the stimulus size triggering avoidance behavior was calculated from the time when the animal started to move in response to the expanding stimulus. To determine the position of stimulus relative to the animal, two parameters, ξ and Φ were calculated. The former parameter represents vertical eccentricity and the latter one represents counterclockwise lateral eccentricity with respect to the sagittal plane. We defined areas A, B and C as follows (Fig.1) and compared the mean angular threshold in each area. Area A: outside an area which has an extent of 18° directly above the animal on a display located at 20 cm, Area B: within an area which has an extent of 18° directly above the animal on a display located at 20 cm, Area C: within an area which has an extent of 18° directly above the animal on a display located at 40 cm. The direction of escape jump was determined from the position of the animal in the first and the last still video frames. Averages reported here are given as the mean ± the standard deviation.

3 Results

The region where the looming stimulus can be presented changes depending on the location of a display. At a distance of 40cm, the stimulus can be presented only within the visual angle corresponding to area C. On the other hand, at 20cm, the stimulus can be presented within both areas B and A (Fig. 1).

If the hypothesis 1 is correct, the threshold in area B should be the same as that in area C, but different from that in area A. On the other hand, if the hypothesis 2 is correct, the threshold in area B should be the same as that in area A, but different from that in area C. The mean angular thresholds in areas A, B and C were 16.8 ± 6.0° (n=39), 15.7 ± 7.4° (n=18) and 12.1 ± 5.4° (n=64), respectively. Statistical analysis (one-way ANOVA and Tukey-Kramer Test) showed that the mean angular threshold in areas A and B were not significantly different ($p>0.05$). The mean angular thresholds in areas B and C were not significantly different, either ($p>0.05$). However, the analysis showed that the mean angular threshold in area C was significantly smaller than that in area A ($p<0.05$). These results suggest that the difference in angular threshold depending on the location of a display is caused by both visual field specificity and integration of size and depth information of an approaching object (Fig.1).

The frog mainly jumps forward to the stimulus in the frontal visual field, on the other hand, the animal, in some cases, turns and jumps backward to the stimulus in the caudal visual field. Statistical analysis (Student's *t*-test) showed that the mean angular thresholds for escape behavior to the stimulus presented in the frontal visual field and the caudal visual field (13.5 ± 6.0° (n=139) and 13.5 ± 6.6° (n=110), respectively) were not significantly different ($p>0.05$). On the other hand, the analysis showed that the mean angular threshold for backward jump was significantly smaller than that for forward jump (11.9 ± 4.3° (n=27) and 13.8 ± 6.5° (n=113), respectively) ($p<0.05$). This suggests that the angular threshold also changes depending on the prospective behavior performed by the animal.

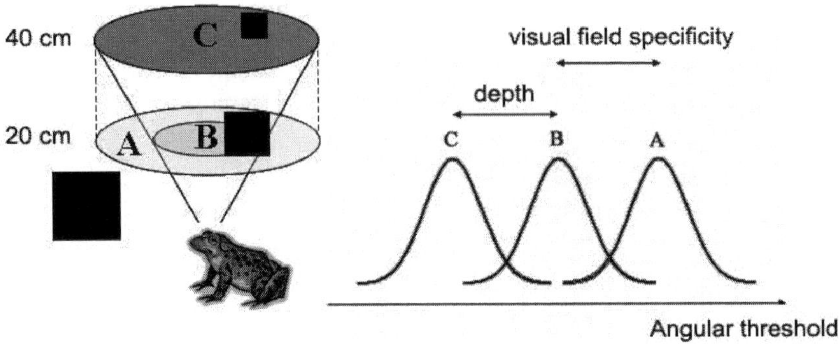

Fig. 1. Angular thresholds in area A, B and C. The mean angular threshold in area C is significantly smaller than that in area A. The difference is caused by both visual field specificity and the integration of size and depth information of an approaching object.

4 Discussion

At first, we discuss adaptive meaning of changing the angular threshold depending on the location of visual stimulus. We showed that the frog processes both retinal size and depth information of an approaching object. The first point should be considered is why the threshold is smaller at the distance of 40 cm than 20cm? The object which reaches a certain angular size far from the animal is bigger than the object which reaches the same size close to the animal. The former object could be more dangerous than the latter one. Therefore, animals should start avoidance behavior earlier for the former stimulus than the latter one. We also showed that angular threshold shows visual field specificity. The next point should be discussed is why the threshold is smaller directly above the animal than in the peripheral regions. The direction of avoidance jump cannot be determined uniquely to the stimulus appearing directly above the animal. Animals could be at a loss where to go and delay the start of response. As a result, animals should start avoidance behavior earlier to the object approaching directly above them.

Second, we consider functional meaning of changing the angular threshold depending on behavioral strategy. We examined the direction of escape jump in detail. In some cases, the frog seems to perform cut-back maneuver as a football player to visual objects approaching in the caudal visual field [1, 2]. The function of cutting-back is easy to imagine, as when a football player perceives a would-be tackler. The smart move is to cut behind the pursuer, who is unable to make the hair-pin turn required to grasp his quarry. If the frog cut back to the stimulus appeared in the caudal visual field, he must turn and jump to the stimulus. Therefore, for successful avoidance, the animal must start to move earlier to compensate delay of avoidance due to turning behavior to the stimulus in the caudal visual field. To this goal, the angular threshold could be smaller in the caudal visual field than the frontal visual field. However, there was no significant difference between them. Here, we should note that the frog does not necessarily make a backward jump to the stimulus in the caudal visual field. Thus, in the next step, we compared the angular threshold between forward and backward jumps. The analysis showed that the mean angular threshold for backward jump was significantly smaller than that for forward jump. This suggests that the angular threshold changes depending on the prospective behavior and could compensate delay of avoidance owing to cut-back maneuver to the stimulus in the caudal visual field.

Acknowledgement

This work was supported by a COE program (center #J19) granted to Kyushu Institute of Technology by MEXT of Japan.

References

1. Ingle, D.: Control of frog evasive direction. In: Arbib, M.A., Ewert, J.P. (eds.) Visual structures and integrated functions. Springer, Heidelberg (1991)
2. Ingle, D.J., Hoff, K.v.S.: Visually elicited evasive behavior in frogs. Bioscience 40, 284–291 (1990)
3. Nakagawa, H., Takemura, Y.: The strategy of collision avoidance behavior of the frog and visual field-specificity of the retinal threshold size. Comp. Biochem. Physiol. 145B, 18 (2005)
4. Robertson, R.M., Johnson, A.G.: Retinal image size triggers obstacle avoidance in flying locusts. Naturwissenschaften 80, 176–178 (1993)
5. Schiff, W.: Perception of impending collision: A study of visually directed avoidance behavior. Psychol. Monogr. 79, 1–26 (1965)
6. Yamamoto, K., Nakagawa, H.: Input and output characteristics of collision avoidance behavior in the frog Rana catesbeiana. Brain Behav. Evol. 62, 201–211 (2003)

Spike-Timing-Dependent LTP/LTD Caused by Uncorrelated Signals through Medial and Lateral Perforant Pathways in the Dentate Granule Cell

Yukihiro Nonaka and Hatsuo Hayashi

Department of Brain Science and Engineering, Graduate School of Life Science and Systems Engineering, Kyushu Institute of Technology, 2-4 Hibikino, Wakamatsu-ku, Kitakyushu 808-0196, Japan
nonaka-yukihiro@edu.brain.kyutech.ac.jp,
hayashi@brain.kyutech.ac.jp

Abstract. The medial and lateral perforant pathways (MPP and LPP) conveying spatial and non-spatial information terminate within the middle and the outer 1/3 of the dentate molecular layer, respectively. MPP projection neurons generate subthreshold oscillations, whereas LPP projection neurons do not, implying that the MPP signal is regular and the LPP signal is rather random. In this paper, we investigated synaptic modification caused by uncorrelated signals through MPP and LPP synapses using a multi-compartmental model of the dentate granule cell. MPP and LPP synapses were subject to an asymmetric spike-timing-dependent plasticity (STDP) rule. As a result, the MPP synaptic weight was enhanced and the LPP synaptic weight was rather depressed. This suggests that LPP signal conveying non-spatial information promotes enhancement of MPP synaptic weight where spatial information passes through.

1 Introduction

The medial perforant path (MPP) originating from the medial entorhinal cortex (MEC) terminates within the middle 1/3 of the dentate molecular layer and the lateral perforant path (LPP) from the lateral entorhinal cortex (LEC) terminates within the outer 1/3 of the molecular layer [1]. MPP and LPP carry spatial and non-spatial information, respectively [2]. Moreover, MPP projection neurons of the MEC generate subthreshold oscillations in synchrony [3], which pace the firing of the MEC neurons, whereas LPP projection neurons of the LEC do not have membrane properties to generate such intrinsic oscillations [4]. This implies that the MPP signal carrying spatial information does not correlate with the LPP signal carrying non-spatial information.

On the other hand, recent study has shown that modification of the LPP synaptic conductance is characterized by an asymmetric spike-timing dependent plasticity (STDP) [5]. Although the learning rule of MPP synapses has not been elucidated, it is possible that MPP synapses are also subject to a STDP rule.

A. Hanazawa et al. (Eds.): Brain-Inspired Info. Technology, SCI 266, pp. 115–118.
springerlink.com

In this study, we investigated synaptic modification caused by uncorrelated signals through MPP and LPP synapses using a multi-compartmental model of the dentate granule cell.

2 Dentate Granule Cell Model and Simulation Protocol

Aradi and Holmes have developed a 47-compartmental model of the dentate granule cell that contains 14 dendrite compartments, two soma compartment and 31 axon compartments [6]. For simplicity, their model was reduced to a 4-compartmental model consisting of three dendrite compartments and one soma compartment (Fig.1). A regular pulse train (interpulse interval = 125 ms) was fed to the middle dendrite compartment through a MPP synapse, and a random pulse train (mean interpulse interval = 100-500 ms, SD = 3-60 ms, Gaussian distribution) was fed to the distal dendrite compartment through a LPP synapse. We assumed that both synapses were subject to an asymmetric STDP rule obtained by Lin and his colleagues [5].

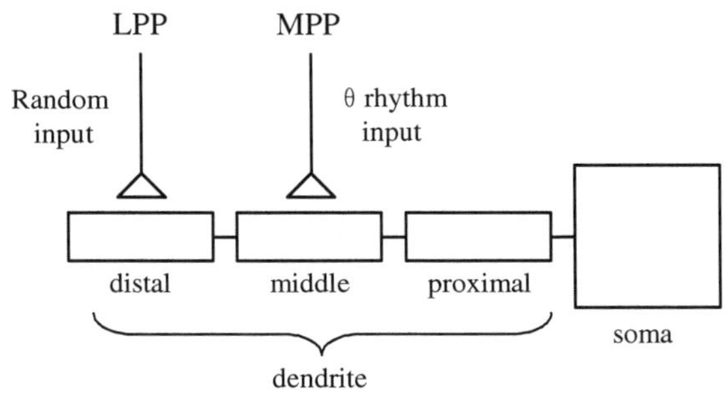

Fig. 1. Four-compartmental model of the dentate granule cell.

The simulation was repeated 5 times for each set of stimulus parameters using different initial conditions. The spike-timing was measured as the time difference between a pulse fed to each synapse and a backpropagating action potential (BAP) passing through the corresponding synapse. The positive spike-timing signified that synaptic activation by the pulse preceded the BAP. The negative spike-timing signified paring in reverse order. Initial conditions of the synaptic conductances were set so that a single EPSP did not fire any action potentials though an integrated EPSP due to almost simultaneous MPP and LPP inputs did fire an action potential.

3 Results

When MPP and LPP synapses were simultaneously stimulated by regular and random pulse trains respectively, both synapses were slightly enhanced at the initial stage (Fig. 2, S1). Then, the MPP synapse was enhanced at a relatively high grow-rate at the second stage, while synaptic conductance of the LPP synapse was maintained or rather depressed (Fig. 2, S2).

The slight enhancement of both synapses at the initial stage was caused by occasional paring of MPP and LPP pulses that produced an integrated EPSP large enough to cross the firing threshold, although MPP and LPP signals did not correlate with each other. In those cases, spike-timings were positive because both EPSPs at MPP and LPP preceded a BAP caused by the integrated EPSP. Moreover, the MPP synapse has the advantage of locating closer to the soma, because spike-latency depends asymmetrically on the order of MPP and LPP pulses. This leads the time difference between EPSP and BAP at the MPP synapse to be statistically smaller than that at the LPP synapse. Consequently, the slight enhancement of the MPP synapse is relatively larger than that of the LPP synapse.

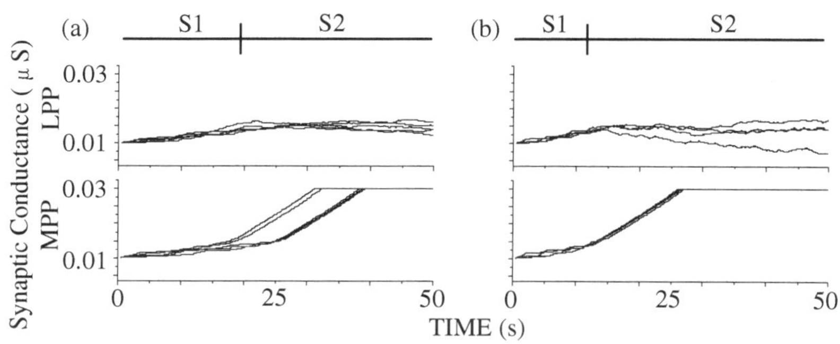

Fig. 2. Synaptic conductances. Mean interpulse intervals of the LPP signal (random pulse train) in (a) and (b) are 250 and 125 ms, respectively. SD = 20ms. The interpulse interval of the MPP signal (regular pulse train) is 125 ms. Five traces obtained using different initial conditions of the granule cell are superimposed.

The asymmetric enhancement of MPP and LPP synapses at the initial stage allowed MPP input to get to elicit an action potential by itself earlier than LPP input. Therefore, the MPP synaptic conductance increased much faster than the LPP synaptic conductance at the second stage. On the other hand, spike-timings at the LPP synapse were consequently distributed from negative to positive because most of the BAPs were induced by regular MPP input and did not correlate with LPP random pulses. Thus, the LPP synaptic conductance was maintained or rather decreased if the STDP rule had the LTD area slightly larger than the LTP area. We also found that the transition point from the initial stage to the second stage was advanced with decrease in the mean interpulse interval of the random pulse train through LPP (Fig. 2a and b). High-frequency random pulses may accelerate enhancement of MPP and LPP synaptic conductances at the initial stage.

4 Summary and Conclusion

When MPP and LPP synapses were simultaneously stimulated by regular and random pulse trains respectively, the MPP synapse was finally enhanced and the

LPP synapse was rather depressed. The present results suggest that MPP synapses where spatial information passes through can be enhanced with the aid of an uncorrelated LPP signal conveying non-spatial information such as odor. This mainly results from asymmetric STDP rules and different distances between those synapses and the soma.

Acknowledgements

This work was supported by a COE program (center #19) granted to Kyushu Institute of Technology by MEXT of Japan and KAKENHI (19500126) granted to one of the authors by JSPS.

References

1. McNaughton, B.L.: Evidence for two physiologically distinct perforant pathways to the fascia dentate. Brain Res. 199, 1–19 (1980)
2. Hargreaves, E.L., Rao, G., Lee, I., Knierim, J.J.: Major dissociation between medial and lateral entorhinal input to dorsal hippocampus. Science 308, 1792–1794 (2005)
3. Dickson, C.T., Magistretti, J., Shalinsky, M., et al.: Oscillatory activity in entorhinal neurons and circuits. Ann. N. Y. Acad. Sci. 911, 127–150 (2000)
4. Wang, X., Lambert, N.A.: Membrane properties of identified lateral and medial perforant pathway projection neurons. Neurosci. 117, 485–492 (2003)
5. Lin, Y.W., Yang, H.W., Wang, H.J., et al.: Spike-timing-dependent plasticity at resting and conditioned lateral perforant path synapses on granule cells in the dentate gyrus: different roles of N-methyl-D-aspartate and group I metabotropic glutamate receptors. Eur. J. Neurosci. 23, 2362–2374 (2006)
6. Aradi, I., Holmes, W.R.: Role of multiple calcium and calcium-dependent conductances in regulation of hippocampal dentate granule cell excitability. J. Comp. Neurosci. 6, 215–235 (1999)

Spatiotemporal Synchronization of Neuronal Activity in a Hippocampal CA3 Network Model Including the O-LM Cell

Seiichi Matsubara, Hatsuo Hayashi, and Kazuki Nakada

Department of Brain Science and Engineering, Graduate School of Life Science and Systems Engineering, Kyushu Institute of Technology, 2-4 Hibikino, Wakamatsu-ku, Kitakyushu 808-0196, Japan
matsubara-seiichi@edu.brain.kyutech.ac.jp,
{hayashi,nakada}@brain.kyutech.ac.jp

Abstract. Neurons often cause bursting discharges in synchrony in the hippocampal CA3 region, which has rich excitatory and inhibitory recurrent connections. It has been supposed that the recurrent inhibition is important to synchronize the neuronal activity. Many types of interneurons exist in the CA3 region. In particular, oriens-lacunosum moleculare (O-LM) cells have a distinctive feature; the cells are resonant and have a resonance frequency in the theta range. In this paper, we investigated how the O-LM cells effectively promoted synchronization of neuronal activity using a hippocampal CA3 network model. As a result, although synchronization of spike bursts of pyramidal cells was improved a little due to the resonant activity of the O-LM cell in comparison with the network with fast- spiking inhibitory neuron, the synchronization was still poor. However, when the basket cells inhibited the O-LM cell unilaterally, synchronization of spike bursts of pyramidal cells was improved dramatically.

1 Introduction

Rich inhibitory recurrent connections exist in the hippocampal CA3 region, besides excitatory recurrent connections. It has been supposed that they play crucial roles in controlling the activity of the CA3 region. Fast-spiking neurons such as basket cells receive excitatory inputs from pyramidal cells and fire high-frequency spikes. Since these interneurons then inhibit pyramidal cells extensively, it has been supposed that they may pace firing of the pyramidal cells [1]. On the other hand, oriens-lacunosum moleculare (O-LM) cells are resonant. Although the role of the O-LM cells in spatiotemporal activity in CA3 remains unclear, it is expected that recurrent inhibition via O-LM cells promotes synchronization of neuronal activity in CA3 because of the resonant activity of O-LM cells in the theta range.

In this paper, using a hippocampal CA3 network model, we investigated promotion of spatiotemporal synchronization due to a cascade of inhibitions via basket and O-LM cells.

A. Hanazawa et al. (Eds.): Brain-Inspired Info. Technology, SCI 266, pp. 119–122.
springerlink.com

2 Methods

The hippocampal CA3 network model consisted of 16×16 pyramidal cells, 25 basket cells, and one O-LM cell (Fig.1). The conductance-base models of the pyramidal and the basket cells developed by Tateno et al. [2] and the O-LM cell developed by Kunec et al. [3] were used. All of them were single-compartment models. Each pyramidal cell was recurrently connected to eight nearby pyramidal cells through excitatory synapses. Connections between pyramidal cells and interneurons will be mentioned in the following section. Although this network model showed the various spatiotemporal activities depending on excitatory and inhibitory connection weights, a set of the weights that caused a theta rhythm was used, because the O-LM cell had a resonance frequency in the theta range.

Fig. 1. Hippocampal CA3 network model. ▲: pyramidal cell. ○:basket cell. ■: O-LM cell.

3 Results

When only basket cells were included in the network as inhibitory interneurons (Fig. 2ai), each basket cell received excitatory inputs from sixteen nearby pyramidal cells and inhibited the same pyramidal cells. In this network, spatiotemporal activity was complex as shown in a raster plot (Fig. 2aii) and the number of firing neurons varied irregularly (Fig. 2aiii) though the field current obtained by summing up synaptic currents of the pyramidal cells in the central area of the network showed a theta rhythm (Fig. 2aiv). Fig. 2av shows firing patterns of a pyramidal and a basket cells.

When an O-LM cell alone was included in the network as an inhibitory interneuron (Fig. 2bi), the O-LM cell received excitatory inputs from all of the pyramidal cells and inhibited them. The degree of synchronization of this network was still poor (Fig. 2bii and biii) and the field current was rather irregular (Fig. 2biv), though synchronization appears to be improved a little due to the resonant activity of the O-LM cell. The poor synchronization resulted from discharges of the interneuron that loosely kept pace with bursts of pyramidal cells (Fig. 2bv).

These firings of the O-LM cell rather disturbed the activity of the pyramidal cells. The network including both basket and O-LM cells showed poor synchronization as well (Fig. 2c).

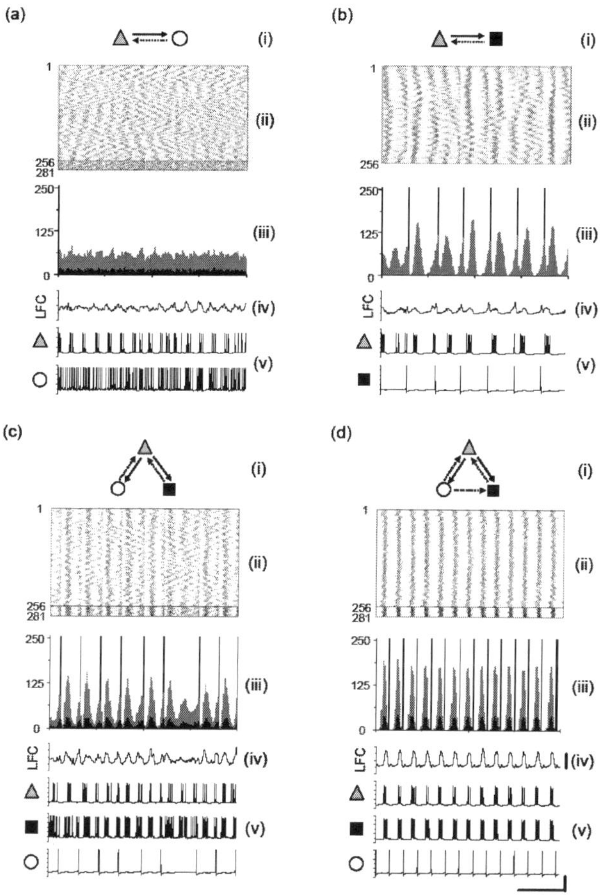

Fig. 2. Patiotemporal activity of the CA3 network model. (i) Connectivity between neurons contained in the network model. (ii) Raster plots. (iii) The number of firing neurons as a function of time. (Gray) Pyramidal cells. (Black) Basket cells. Vertical lines indicate firing of the O-LM cell. (iv) Local field current. (v) Voltage traces of individual neurons in this model. Vertical bar: (iv) 1.0nA, (v) 60mV. Horizontal bar: 500ms.

In contrast, when there existed unilateral inhibitions from basket cells to the O-LM cell in addition to the above recurrent inhibitions, synchronization of bursts of pyramidal cells was improved dramatically (Fig. 2dii and 2diii). Firing of pyramidal and basket cells were well clustered and periodic. Field current is also periodic (Fig. 2div). As shown in Fig. 2dv, the O-LM cell fired just after the spike bursts of pyramidal cells. In other words, the O-LM cell was inhibited by burst firing of the basket cells accompanying the bursts of the pyramidal cells and then fired rebound spikes after the firing of the basket cells terminated. As a result, the O-LM cell firing hyperpolarized pyramidal cells efficiently and paced the next burst firing of the pyramidal cells, resulting in synchronization.

4 Summary and Conclusion

Recurrent inhibitions of pyramidal cells via single interneurons, i.e. basket or O-LM cells, were insufficient to synchronize bursts of the pyramidal cells. However, unilateral inhibitions from basket cells to the O-LM cell employed in addition to the above recurrent inhibition synchronized bursts of the pyramidal cells dramatically. These results suggest that the cascade of the inhibitions via basket and O-LM cells do not interfere with bursts of the pyramidal cells but are effective to synchronize the bursts.

Acknowledgement

This work was supported by a COE program (center #J19) granted to Kyushu Institute of Technology by MEXT of Japan and KAKENHI (19500126) granted to one of the authors by JSPS.

References

1. Cobb, S.R., Buhl, E.H., Halasy, K., et al.: Synchronization of neuronal activity in hippocampus by individual GABAergic interneurons. Nature 378, 75–78 (1995)
2. Tateno, K., Hayashi, H., Ishizuka, S.: Complexity of spatiotemporal activity of a neural network model which depends on the degree of synchronization. Neural Networks 11, 985–1003 (1998)
3. Kunec, S., Hasselmo, M.E., Kopell, N.: Encoding and Retrieval in the CA3 Region of the Hippocampus: A Model of Theta-Phase Separation. J. Neurophysiol. 94, 70–82 (2005)

The Regulation of Glutamate-Induced Astrocytic $[Ca^{2+}]i$ Responses by the Desensitization of Metabotropic Glutamate Receptor

Isao Goto and Kiyohisa Natsume

Department of Brain Science and Engineering, Graduate School of Life Science and Systems Engineering, Kyushu Institute of Technology, 2-4 Hibikino, Wakamatsu, Kitakyushu 808-0196, Japan
goto@brain.kyutech.ac.jp

Abstract. Astrocyte (AS) has metabotropic glutamate receptor type 5 (mGluR5) and forms tripartite synapse with neurons. AS is thought to receive the information from neurons and to play an important role of information processing in the CNS. In hippocampal cultured AS, the application of glutamate induces the four patterns of astrocytic Ca^{2+} responses. They were a sustained oscillation, a damped oscillation and a step-rise response. The astrocytes have the varieties of the intracellular Ca^{2+} responses. Why does the variety come from? In the present study, we propose the desensitization process of mGluR5 causes the varieties of the astrocytic responses.

1 Introduction

Astrocytes locate around the synapse, and construct "Tripartite synapse" with presynaptic cell and postsynaptic cell [1]. There are some reports that astrocytes may interact with neurons via some molecule [2]. On the other hand, the experimental results revealed that cultured astrocytic Ca^{2+} responses to the glutamate ([Glu]o) had the four patterns [3]. These patterns are oscillation, oscillation that frequency gradually decreased, a damped oscillation and step response. The four patterns emerge even in the same Glu concentration. Thus astrocytes have heterogeneity of $[Ca^{2+}]i$ responses. Such heterogeneity may be useful for the information processing. However it is not known how heterogeneity is induced. We thought that the heterogeneity might be evoked via desensitization of metabotopic Glu receptor type 5 (mGluR5). In this study, to investigate this idea, we constructed a simple model that considered the desensitization process of mGluR5, and could simulate the heterogeneity of astrocytic $[Ca^{2+}]i$ responses against Glu stimulation. Moreover we then verified the validity of this model on the other properties. These results suggest where the heterogeneity of astrocytic responses comes from.

A. Hanazawa et al. (Eds.): Brain-Inspired Info. Technology, SCI 266, pp. 123–127.
springerlink.com

2 Model

2.1 PLCδ Model (Modified Young and Keizer Model)

In hippocampal CA1 area the neuroral firing release Glu from the presynaptic terminal. The released Glu binds not only to glutamate receptor on the postsynaptic membrane but also to mGluR5 on astrocytic membrane. The mGluR5 is activated by Glu, and is also regulated by intracellular protein kinase C (PKC) and protein phosphatase (PP). The activated mGluR5 induces hydrolysis of phosphatidyl inositol bisphosphate (PIP_2) by phospholipase C (PLC) β via G-protein activation. The hydrolysis produces inositol trisphosphate (IP_3) that binds to and opens the IP_3 receptor/ Ca^{2+} channel (IP_3R) of endoplasmic reticulum (ER) Ca^{2+} store. When IP_3R is activated, then Ca^{2+} is released into intracellular space. In PLCδ model, which we proposed previously, $[Ca^{2+}]i$ can activate PLCδ that also hydrolyzes PIP_2 and that produces IP_3 again. These processes were included into PLCδ model (Fig. 1). The PLCδ model is modified from De Young and Keizer's model [4]. The details of the PLCδ model were described in the previous paper [5].

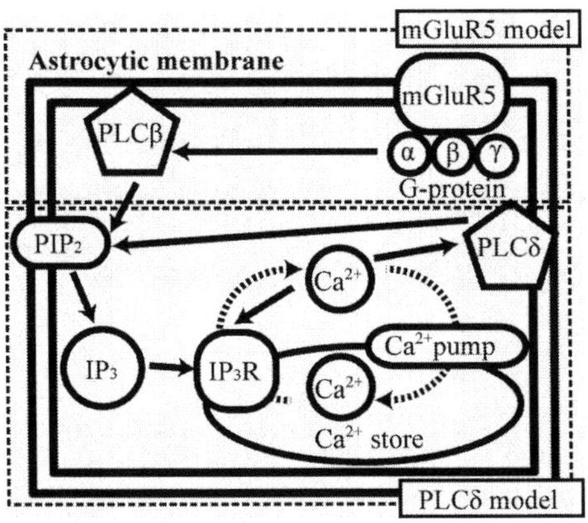

Fig. 1. mGluR5and PLCδ model (MP model): mGluR5: Metabotropic glutamate receptor, PLC: Phospholipase C, PIP_2: Phosphatidylinositol bisphosphate, IP_3: Inositol trisphosphate, IP_3R: Inositol trisphosphate receptor

2.2 mGluR5 Desensitization Model (MD Model)

mGluR5 is desensitized by the phosphrylation of the receptor by PKC[6]. The receptor is resensitized by the dephosphorylation by PP. We assumed that in the desensitization process, there are four states of mGluR5 (Fig. 2: R, R_Glu, D_R_Glu and D_R). When mGluR5 (R) is stimulated by Glu, mGluR5 binds with Glu. R_Glu is the complex of mGluR5 and Glu, the complex is phosphorylated by PKC and the receptor is desensitized. Here, we assumed that the desensitization occurred only when the receptor is in the state R_Glu and that in the desensitization states (D_R_Glu), Glu rapidly dissociates from D_R_Glu and the receptor becomes D_R form. The form of D_R does not bind with Glu. Because of this assumption, D_R_Glu does not return to R_Glu. Only D_R returns to R state. The desensitization and resensitization of mGluR are corresponding to the phosphorylation by PKC and dephosphorylation by PP of the receptor. The rates of desensitization (***kf_2***) and resensitization (***kf_4***) are corresponding to the rate of phosphorylation and dephosphorylation, respectively. MD model also simulated the reaction of GDP/GTP exchange. mGluR5 binds with Glu, R_Glu binds with G-protein, GDP attached to G-protein is exchanged with GTP and G-protein becomes an activated form. The activated G-protein activates PLCβ and IP_3 is generated. The reactions in downstream from here can be simulated in our previous PLCδ model [5].

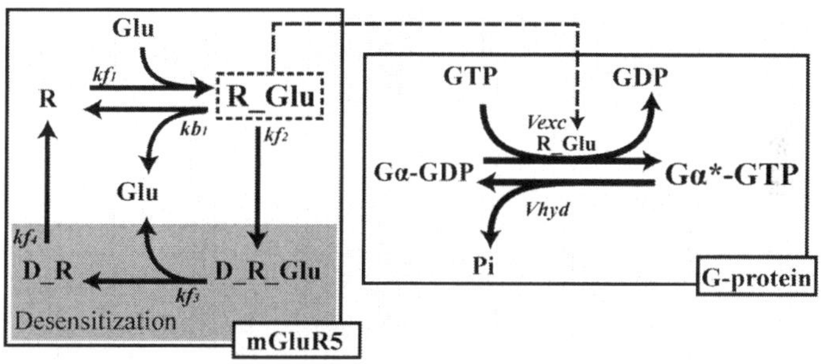

Fig. 2. mGluR5 desensitization processes and G-protein activation processes (mGluR5 model)

2.3 A Simple [Ca^{2+}]iresponse Model (MP Model)

Here, we propose MP model including PLCδ and MD model.

3 Results

We first reproduced the Glu-induced heterogeneity observed in cultured astrocytes [3]. In 100 μM Glu application, cultured astrocytes showed various pattern of [Ca^{2+}]i responses [3]. MP model induced heterogeneity with change in the mGluR5 desensitization rate ***kf_2*** (Fig. 2A-D). As shown in the figures, temporal decrease of R_Glu was induced with the increase in ***kf_2***. The response patterns of AS[Ca^{2+}]i were regulated by ***kf_2***. We first set ***kf_2*** = 0.0001 /sec, and represent damped oscillation of AS[Ca^{2+}]i responses (Fig. 2A). Then, ***kf_2*** was increased, and regular oscillation and the oscillation with step-rise responses were induced (Fig. 2B-D).

We also calculated the period of astrocytic [Ca^{2+}]i oscillation induced by different concentration of [Glu]o (Fig. 4A). The periods decreased with the increase in [Glu]o. We also calculated the cell-responding ratio in the four [Glu]o induced

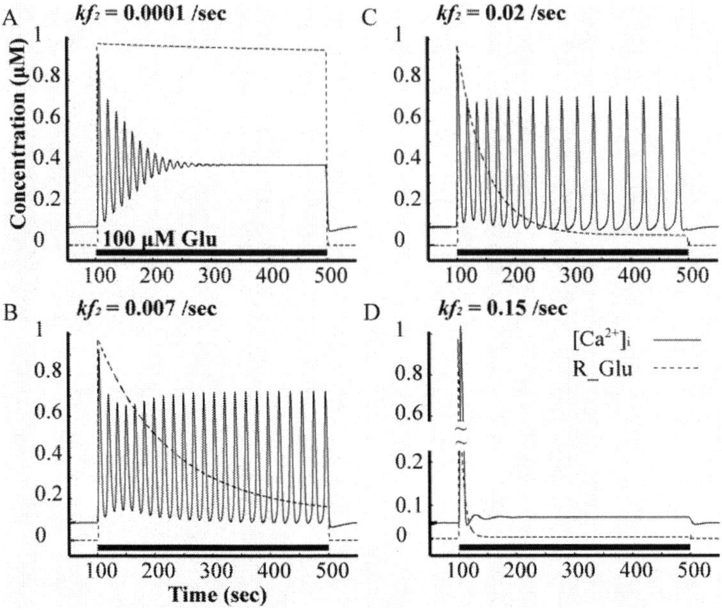

Fig. 3. A-D. Glu-induced AS[Ca^{2+}]i responses.
The model reproduced all four response patterns reported by [3]. When Glu was applied at 100 μM for 400 sec (thick black bar), the responses were induced by changing kf_2, the rate of desensitization. A. a damped oscillation (kf_2 = 0.0001 /sec). B. a sustained oscillation (kf_2 = 0.007 /sec). C. a sustained oscillation, with frequency gradually decreasing (kf_2= 0.02 /sec). D. a step-rise response (kf_2 = 0.15 /sec).

[Ca^{2+}]i patterns. We assumed that astrocytic PLCβ and PLCδ concentration may differ in each astrocytes. Then, we distributed uniformly these enzymes concentration in the limited area and calculated the ratio of [Ca^{2+}]i responses (cell-responding rate) against each [Glu]o. The ratin increased with the increase in [Glu]o (Fig. 4B). These simulations showed good agreement with the experimental results [3].

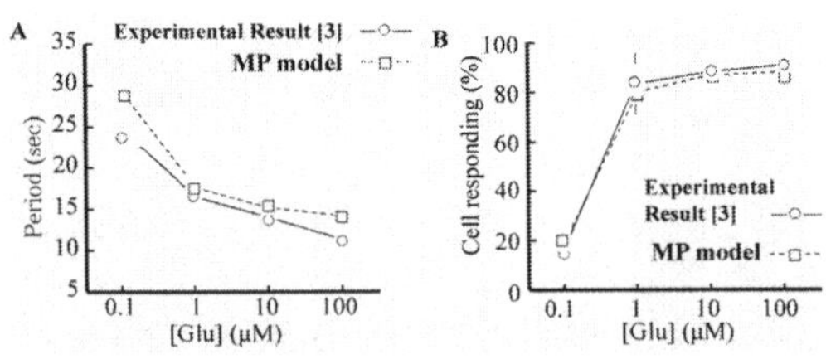

Fig. 4. A. Comparison of period of [Ca^{2+}]i oscillation with the experimental results[3]. B. Comparison of the ratio of responded cell number to the total number with the experimental results[3]. A and B show the good agreement between the results of simulations and experiments.

4 Discussion and Conclusion

Our MP model could induce the various response patterns and the phenomena obtained in the cultured astrocytes [3].

Recent study suggests that mGluR5 desensitization rate was regulated by PKC [7]. Fig. 3A-D suggested that astrocytic heterogeneity of $[Ca^{2+}]i$ responses to Glu stimulation may come from the difference in the desensitization. The results suggested that the heterogeneity can be rate perhaps due to different concentrations of intracellular PKC.

Glu stimulation induces IP_3 production. From our previous study, the periods of $AS[Ca^{2+}]i$ oscillation were induced by increasing IP_3. Thus increasing [Glu]o caused the decreasing periods of $AS[Ca^{2+}]i$ oscillation (Fig. 4A).

A simple description of mGluR5 desensitization process by state transition scheme allows our model for $AS[Ca^{2+}]i$ oscillation to be simple, while it shows a good quantitative agreement with the experimental results. Our new model would help to build AS-Neuron network model in the brain.

Acknowledgement

This work was partially supported by the 21th Century COE (Center of Excellence) Program in Kyushu Institute of Technology, entitled "World of brain computing interwoven out of animals and robots".

References

1. Kudo, Y.: SAIBOUKOUGAKU 22, 390–392 (2003)
2. Bezzi, P., et al.: Nature 391, 281–285 (1998)
3. Cornell-Bell, A.H., Finkbeiner, S.M., Cooper, M.S., Stephan: Science 247, 470–473 (1990)
4. De Young, G.W., Keizer, J.: Proc. Natl. Acad. Sci. 89, 9895–9899 (1992)
5. Goto, I., Kinoshita, S., Natsume, K.: Neurocomputing 58, 461–467 (2004)
6. Balázs, R., et al.: J. Neurochem. 69, 151–163 (1997)
7. Francesconi, A., Duvoisin, R.: PNAS 97, 6185–6190 (2000)

Functional Properties of Resonate-and-Fire Neuron Circuits for Bio-Inspired Chemical Sensor Array

Kazuki Nakada, Katsumi Tateno, Hatsuo Hayashi, and Kiyonori Yoshii

Department of Brain Science and Engineering, Graduate School of Life Science and Systems Engineering, Kyushu Institute of Technology, 2-4 Hibikino, Wakamatsu, Kitakyushu 808-0196, Japan
{nakada,tateno,hayashi,yoshii}@brain.kyutech.ac.jp

Abstract. The purpose of our project is to develop brain-inspired chemical sensor arrays from physiological, theoretical, and engineering points of view. In the previous work, a computational model for chemical sensor arrays has been proposed based on physiological properties of mouse taste bud cells (TBCs). The model consists of three functional parts: the chemical sensor, the random pulse generator, and the stochastic synchronizer. The chemical sensor array based on the computational model detects the concentration of chemical substances as the degree of stochastic synchronization. For practical implementation, we compare the probabilistic firing properties of an array of resonate-and-fire neuron (RFN) models and those of an array of leaky integrate-and-fire neuron models in order to consider their feasibility as the second part of a chemical sensor array. We further investigate stochastic synchronization in uncoupled integrated circuits implementing the RFN model in order to confirm if they act as the third part of a chemical sensor array on a practical hardware platform.

1 Introduction

Towards the development of brain-inspired chemical sensors, we focus on the functional network of mouse taste bud cells (TBCs) in taste information processing, which is a stochastic process due to probabilistic neurotransmitter release and action potential generation with fluctuation [Ohtubo et al.]. In the previous work, a computational model for a chemical sensor array has been proposed based on the physiological properties of the TBCs [Tateno et al.]. The model consists of three functional parts: (i) the chemical sensor part composed of an array of transducers of the concentration of chemical substances to analog signals, (ii) the random pulse generator part with an array of analog-to-pulse converters, each of which is connected to one of the transducers, and (iii) the stochastic synchronizer part that receives a sequence of random pulses as the output of the second part and takes the concentration of chemical substances as the degree of stochastic synchronization (Fig. 1). In practice, the computational model has been implemented on a software platform with commercially available alcohol sensors as a chemical sensor array [Igarashi et al.].

A. Hanazawa et al. (Eds.): Brain-Inspired Info. Technology, SCI 266, pp. 129–133.
springerlink.com

For practical implementation of chemical sensor arrays, it is the demand to guarantee stable operation under constraint conditions: probabilistic behavior of a limited number of sensing elements with nonuniformity. Surprisingly, biological chemical sensor arrays, such as the TBCs, have solutions to fulfill the demand. By learning from the solutions, we consider essential requirements of the functional parts of the chemical sensor array. This will provide valuable insight into how to design each functional part in practical implementation.

In this context, we compare the probabilistic firing properties of an array of the resonate-and-fire neuron (RFN) models [Izhikevich et al.] and those of an array of the leaky integrate-and-fire neuron models. This purpose is to confirm their feasibility as constituting elements for the random pulse generator part. We further investigate stochastic synchronization in an array of uncoupled integrated circuits implementing the RFN model [Nakada et al.] in order to confirm if it act as the stochastic synchronizer part on a hardware platform.

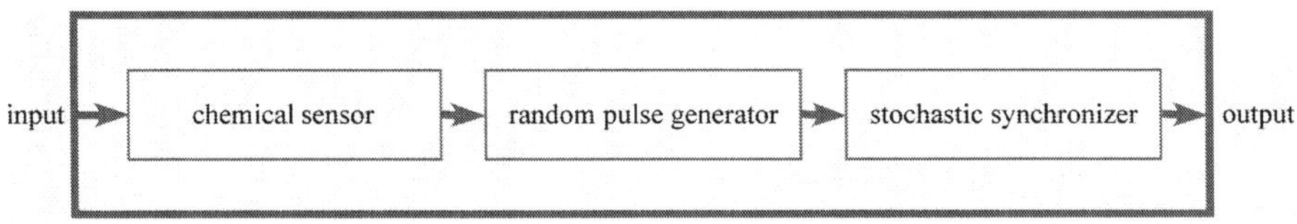

Fig. 1. Block diagram of the computational model for chemical sensor arrays.

2 Constituting Elements for Random Pulse Generator

We here briefly discuss about the difference in the probabilistic firing properties of an array of leaky integrate-and-fire neuron models and those of an array of RFN models in order to confirm their feasibility as constituting elements for the random pulse generator part. We assume that each element is driven by Gaussian white noise as a common input, and has an individual threshold with a standard normal distribution and no connection between elements.

Under this assumption, the interspike interval (ISI) probability density of an array of the leaky integrate-and-fire models is required at most a few tens of elements to be close to an exponential distribution, that is, the firing timing of the array can be approximately regarded as a Poisson process. In contrast, the ISI probability density of an array of a limited number of the RFN models fails to be an exponential distribution. This difference is caused by the probabilistic firing properties of each element owing to the intrinsic characteristics of the membrane potential as driven by Gaussian white noise [Verechtchaguina et al.], leading to distinct functions for the random pulse generator. More specifically, an array of a limited number of leaky integrate-and-fire neuron models can convert the mean of Gaussian white noise as a common input into the mean of ISIs at most a few tens of elements. This characteristic is suitable for constructing an analog-to-pulse converter as a random pulse generator part for a chemical sensor array.

3 Circuit Implementation of Stochastic Synchronizer

We have designed an analog integrated circuit implementing the RFN model in our previous work [Nakada et al.]. The RFN circuit consists of a membrane circuit, threshold-and-fire circuit, and current-mirror integrators as excitatory and inhibitory synaptic circuits. The membrane circuit has second-order dynamics containing two state variables, U_i as a current-like state variable and V_i as a voltage-like state variable. Therefore, the circuit has a voltage-dependent resonant frequency at the resting state, resulting in functional behavior such as coincidence detection, frequency preference, and post-inhibitory rebound, as in the case of biological neurons.

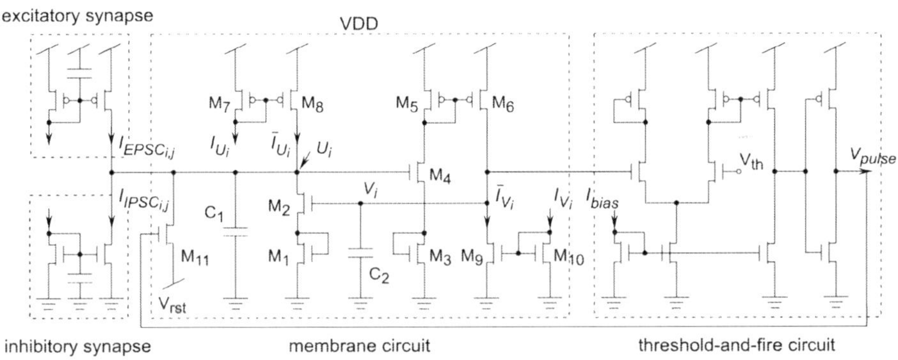

Fig. 2. Schematic diagram of the resonate-and-fire neuron circuit that consists of a membrane circuit, threshold-and-fire circuit, and current-mirror integrators as excitatory and inhibitory synaptic circuits. The membrane circuit has second-order dynamics containing two state variables U_i (a current-like state variable) and V_i (a voltage-like state variable).

We simulated an array of uncoupled RFN circuits with the circuit simulator, SPICE (Simulation Program with Integrated Circuit Emphasis). The bias currents with standard normal distribution were fed into each circuit to have an individual resting potential and an individual resonant frequency. We set the mean and the standard deviation of the bias currents at 10 nA, and 0.5 nA, respectively.

Figures 3A and B show the waveforms of the membrane potential of the RFN circuits, in which we gave common random pulses with Gamma-distributed interpulse intervals (IPIs) into the array of the circuits. The degree of stochastic synchronization was increased with the increase of the coefficient of variation of the IPIs even in the presence of non-uniformity of the circuits. This may arise from the difference in the degree of after hyper-polarization of the circuits. We evaluate the degree of thee stochastic synchronization as the cross correlation of the waveforms of the membrane potential, as shown in Fig. 3C.

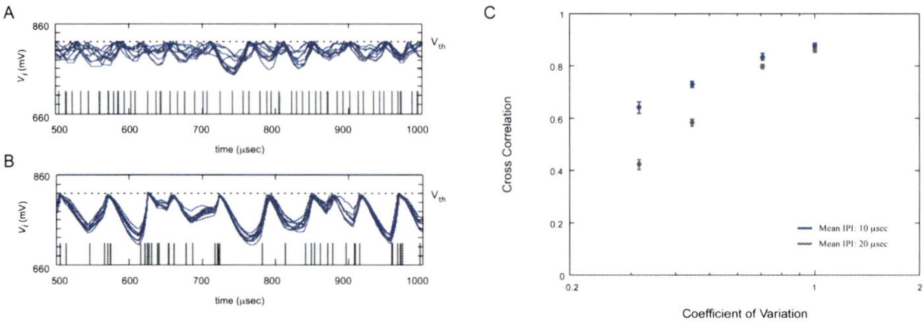

Fig. 3. Stochastic synchronization in the array of non-identical uncoupled resonate-and-fire neuron circuits (the number of the circuits equals to 10) driven by random pulses with Gamma-distributed interpulse intervals (IPIs), in which the coefficient of variation of random pulses equal to (A) $1/\sqrt{10}$ and (B) 1, respectively. (C) The degree of stochastic synchronization, which is represented as the cross correlation of the waveforms of the membrane potential v_i, was increased with the increase of the coefficient of variation of IPIs.

4 Summary

In biological sensory coding, a population of sensory cells may effectively use correlations between these cells in order to transmit information even in a noisy environment. Nonlinear phenomena such as stochastic synchronization increase correlations between uncoupled elements. In fact, the computational model for chemical sensor arrays effectively uses such a phenomenon to reduce the effect of non-uniformity of sensing elements due to device deviation [Tateno et al.].

For practical implementation, we have compared the probabilistic firing properties of an array of the leaky integrate-and-fire neuron models and an array of the resonate-and-fire neuron (RFN) models, it is suggested that the difference in the probabilistic firing properties will provide distinct functions for the random pulse generator part for chemical sensor arrays. In particular, an array of a limited number of leaky integrate-and-fire neuron models can act as an analog-to-pulse converter. This characteristic is suitable for the random pulse generator part for the chemical sensor arrays.

From the circuit simulations using SPICE, the degree of stochastic synchronization in the array of uncoupled RFN circuits is profoundly affected by the coefficients of variation of the random pulses. The results indicate that the array of the RFN circuits can be used as a stochastic synchronizer for chemical sensor arrays in hardware implementation.

Acknowledgements

The authors would like to thank Igarashi J, Ohtubo Y, and Miki T for their fruitful discussions and helpful comments. This work was supported by a COE program (center #J19) granted to Kyushu Institute of Technology by MEXT of Japan.

References

1. Igarashi, J., Tateno, K., Nakada, K., Miki, T., Ohtubo, Y., Yoshii, K.: Presented at ICONIP/BrainIT 2007 (2007)
2. Izhikevich, E.M.: Dynamical Systems in Neuroscience: The Geometry of Excitability and Bursting. MIT Press, Cambridge (2007)
3. Nakada, K., Asai, T., Hayashi, H.: International Journal of Neural Systems 16(6), 445–456 (2006)
4. Ohtubo, Y., Suemitsu, T., Shiobara, S., Matsumoto, T., Kumazawa, T., Yoshii, K.: Journal of Physiology 530, 287–293 (2001)
5. Tateno, K., Yoshii, K., Ohtubo, Y., Miki, T.: International Congress Series 1301, 52–55 (2007)
6. Verechtchaguina, T., Sokolov, I.M., Schimansky-Geier, L.: Biosystems 89(1-3), 63–68 (2007)

A Robot Vision System Using a Silicon Retina

Atsushi Sanada[1,*], Kazuo Ishii[1], and Tetsuya Yagi[2]

[1] Department of Brain Science and Engineering, Graduate School of Life Science and Systems Engineering, Kyushu Institute of Technology, 2-4 Hibikino, Wakamatsu, Kitakyushu 808-0196, Japan
sanada-atsushi@edu.brain.kyutech.ac.jp,
ishii@brain.kyutech.ac.jp

[2] Division of Electrical, Electronic and Information Engineering, Graduate School of Engineering Osaka University, 2-1 Yamadaoka, Suita, Osaka, 565-0871, Japan
yagi@eei.eng.osaka-u.ac.jp

Abstract. In this research, a robot vision system using the 2 pair of rails is developed. This robot vision system is considered use for robot vision or monitoring camera. The silicon retina is developed in Osaka University and is adopted as the visual sensor in this research. The silicon retina has features such as high-speed, low power consumption.

1 Introduction

We can discriminate a certain object among a lot of objects using visual information, whether the object is in the sunlight or under moonlight. Our eyes can adjust the sensitivity over wide range by the dark adaptation or the light adaptation.

The vision chip that imitates biological sight is developed by M. A. Mahowald and C. Mead in 1989 takes attentions because the chip had advantage mentioned above [1]. The vision chip that imitates the circuit structure and the function of the sight system of the living body achieved by an analog CMOS VLSI is especially called a silicon retina. Vision chips had problems such as low sensitivity and much noise. Therefore, vision chips were not used in engineering fields like a robot vision. To solve those problems, a buffer technique for noise compensation is introduced into the vision chip by Yagi et. al. [2][3]. The vision chip has features such as high-speed, low power consumption and can be used in the place where the light intensity changes. This feature is useful for developing the robot and monitoring camera. Robots need advanced sensors like a human eye to adapt to the environment and chase a target and so on.

The target of our research is to develop a stand-alone robot vision system using the silicon retina vision chip.

2 Robot Vision System

2.1 Mechanism

Figure 1 shows the 3D CAD sketch of the robot vision system and the developed robot vision system is shown in Fig. 2. The robot vision system with a silicon

* Corresponding author.

A. Hanazawa et al. (Eds.): Brain-Inspired Info. Technology, SCI 266, pp. 135–139.
springerlink.com

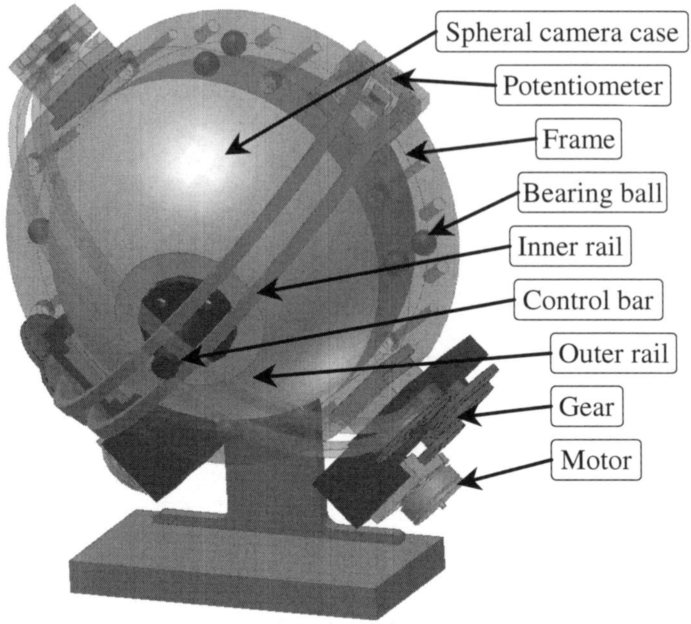

Fig. 1. 3D CAD sketch of robot vision system (Back view)

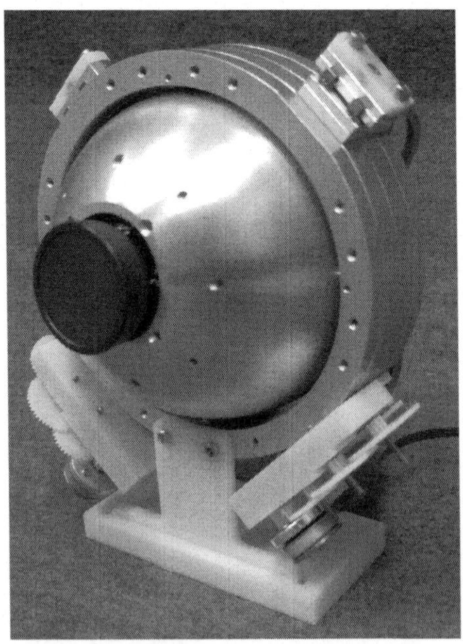

Fig. 2. Developed robot vision system (Front view)

retina has 2 DOF movements. Regarding the 2 DOF motion mechanisms, stepping or DC motors are usually attached to the actuators joints directly. These mechanisms have problems which need large torque to work because the outside motor rotates the camera together with inside actuators. To solve this problem, the mechanism should be simple, small and easily controlled.

We propose a new mechanism that has two kinds of rails. This mechanism is composed of the silicon retina camera, a spheral camera case, a frame, eight bearing balls, two pair of the rails and a control bar. First of all, the silicon retina camera is covered by a spheral camera case. The spheral camera case is located on the center of the frame. The spheral camera case is supported by eight bearing balls which are mounted inside of the frame. Due to the bearing balls, the spheral camera case can rotate smoothly like a ball bearing. On the back of the spheral camera case, the control bar is installed, and the control bar is driven by a pair of the rails. The shape of the control bar which contacts to the outer rail is cylindrical. And the control bar which contacts to the inner rail has a flat surface. Due to the flat surface, the spheral camera case can not rotate in the roll direction.

2.2 System Architecture

Figure 3 shows the system architecture of the vision module. The camera angle is decided by commands from the PC using RS232 communication. The command is sent to a microcomputer (PIC16F88). The microcomputer determines rail angles which can detect by potentiometers using internal A/D converter. The signal of rail rotation is sent to another microcomputer (PIC16F628A) which rotates the rail to the target direction.

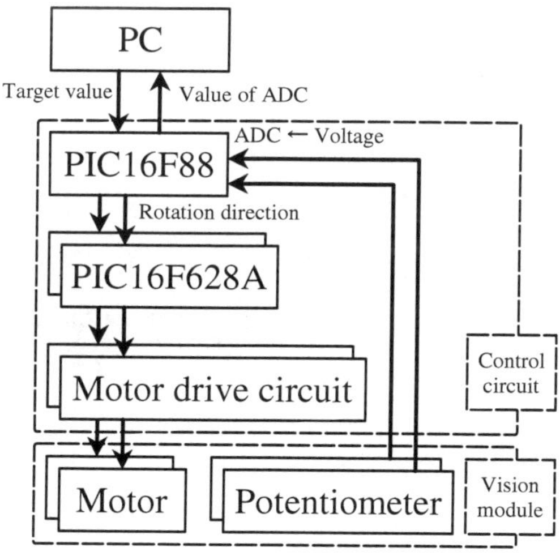

Fig. 3. System architecture

3 Experiments

To evaluate the developed mechanism, the tests to capture the marker are carried out. Fig. 4 shows test environment. The marker is pasted on the wall of 30 centimeters forward of the robot vision system. The number is printed on the marker. The marker is arranged to the circle of 60 cm in the diameter that centers on the marker of number 5. Fig. 5 shows the captured image by the silicon camera. The numbers in the figure shows the angle of inner rail and outer rail. In the test of the capture image, it was confirmed that the robot vision system can follow the instruction and can get the image.

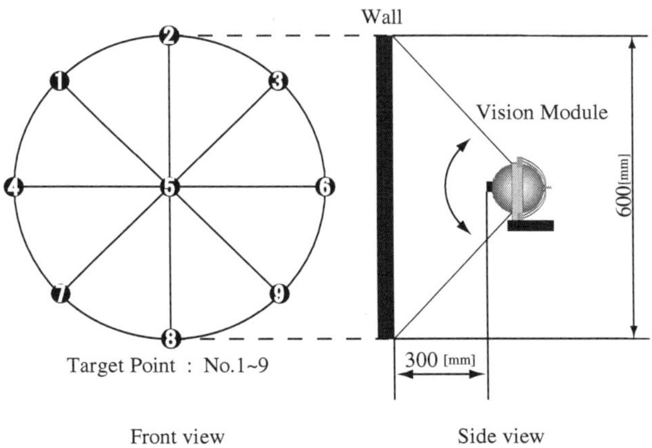

Fig. 4. Test environment

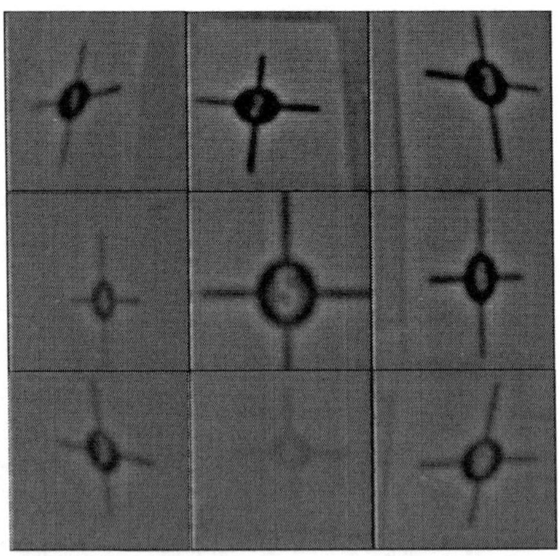

Fig. 5. Captured image

4 Conclusions

In this paper, a prototype of the robot vision system is developed. The system includes a silicon retina and a mechanism that has 2 guide rails. In the tests of image capture, we can see that the robot vision system can follow the instruction and can get the image. This robot vision system will be made smaller in the future.

Acknowledgement

This research was supported by Fukuoka Coal Mining Area Development Center.

References

1. Mead, C.: Analog VLSI and Neural Systems. Translate Shiro Usui, Hiroo Yonedu, Toppan (1993)
2. Yagi, T., et al.: A Parallel Analog Intelligent Vision Sensor with a Variable Receptive Field. The transactions of the Institute of Electronics, Information and Communication Engineers J81-D-I(2), 104–113 (1998)
3. Kameda, S., Yagi, T.: An analog VLSI chip emulating sustained and transient response channels of the vertebrate retina. IEEE Transactions Neural Networks 14(5), 1405–1412 (2003)

Real-Time Human-Machine Interaction System Based on Face Authentication and Arm Posture Recognition

Ishtiaq R. Khan, Takashi Morie, Hiroyuki Miyamoto, Yasutaka Kuriya, and Masaki Shimizu

Department of Brain Science and Engineering, Graduate School of Life Science and Systems Engineering, Kyushu Institute of Technology, 2-4 Hibikino, Wakamatsu, Kitakyushu 808- 0196, Japan
khan@brain.kyutech.ac.jp

Abstract. In this article we present a system for secure human-machine interaction. User verification is done based on face matching, and commands to machine are delivered by arm gesture. Algorithms used in our system for face matching and arm posture detection get some inspiration from brain visual functions and architecture. Higher complexity modules are implemented in hardware to achieve higher computational efficiency needed for real time processing.

1 Introduction

In today's world, interaction of humans with machines has become a matter of routine in our everyday life. In these applications involving human-machine interaction, user verification and reliable communication between users and machines are important issues. User verification is required to restrict the machine to respond only to the authorized user(s), and not to a stranger. Similarly it is required that the user should be able to convey his/her commands to the machine in a natural and reliable way. Like we humans can recognize each other from our faces, it is desirable that machines also do it the same way. In this research, user verification is based on face recognition. Gesture is a natural way of mutual communication among humans, and in this research we propose an arm gesture recognition algorithm, which can be used to convey certain commands from user to the machine.

2 User Verification

Different steps involved in our user verification method are described below:

2.1 Face Detection

Human faces are detected in the input images using the OpenCV's [1] face recognition algorithm. This is an implementation of Viola and Jones method

A. Hanazawa et al. (Eds.): Brain-Inspired Info. Technology, SCI 266, pp. 141–145.
springerlink.com

presented in [6]. Their classifier uses a small number of critical Haar-like features which are evaluated very efficiently using the concept of integral image. The algorithm keeps focus of attention on more promising regions and performs the detection of faces quite efficiently, which makes it suitable for real time applications. The detected faces in our system are mapped to a size of 64 x 64 pixels.

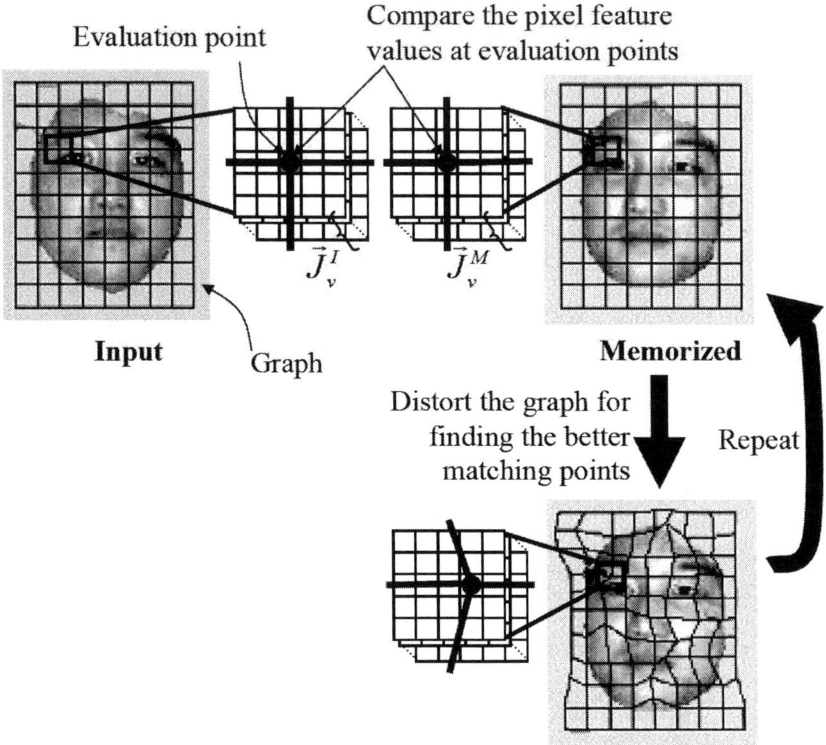

Fig. 1. Elastic Graph Matching (EGM) principle

2.2 Gabor Wavelet Transform

Gabor Wavelet Transform (GWT) is a powerful feature extraction method, which can be used for reliable face/object recognition [2]. Our system keeps a database of the features extracted from faces of the registered users. These features are the GWT coefficients calculated in four directions at five different frequencies, calculated for each pixel on the face. When our system detects a face in proximity, its features are extracted and matched against the database. Calculation of GWT coefficients is a time consuming procedure, and we have implemented it on a dedicated LSI [3]. Our implementation allows a tradeoff between speed and accuracy based on size of grid on which GWT coefficients are calculated for the detected faces.

2.3 Elastic Graph Matching

For comparison of the GWT features of two faces, the principle of elastic graph matching (EGM) [2] is used. In this approach, the features at an evaluation point in one template are matched in a small neighborhood of that point in the other template, and the point is moved to the location of the best match. The distortion caused by movement of all points in the input template is assessed against the quality of matching, and a tradeoff between minimum distortion and better matching is used to determine the best matching face in the stored database. The working of EGM is shown in Fig. 1 for a better understanding. For higher computational efficiency, EGM process has been implemented on an FPGA [4], which can perform the matching of two 100 x 100 images in around 1.0 ms.

3 User-Machine Interactions

If the user face is matched with a face in the database, it is permitted to have further interaction with the machine. In our system, we use arm posture of the authorized user to post certain commands to a robot.

3.1 User-Machine Interactions

The procedure of user verification described above does not use stereo vision, and can in fact be done with a single camera. However, our arm posture recognition algorithm uses stereo vision and needs images captured by both cameras. First, the disparity of the detected face positions in both images is calculated. This is used to define search regions in both images in which the arm can be located. These search regions are further refined by removing background objects, which are considerably far from the cameras compared to the distance of the detected faces from the cameras.

3.2 Arm Model Fitting

A three-dimensional four degrees of freedom human arm model is constructed, and the best matches for its two-dimensional mappings on both images are determined [5]. The arm model is divided into three parts — hand, upper arm and lower arm, and different procedures are used for detection of each part. Hand position is found based on skin color and edge based matching. Our skin color model is not a fixed pre-defined model; rather it adapts to skin color of the user based on color of the matched faces in the current frame. In edge based matching, both magnitude and orientation of the edges are considered. The upper and lower arms are further divided into smaller center and side fins. Center fins detect the center of the arm using SAD (Sum of Absolute Difference) based pixel matching, while side fins detect the contour of the human arm using edge based matching. Suitable weights are assigned to each detected fin to evaluate the best combination of fins, to be recognized as the arm posture. It should be noted that detection of each part of the arm is done independently, and therefore error in detection of one part does not affect the detection in other parts.

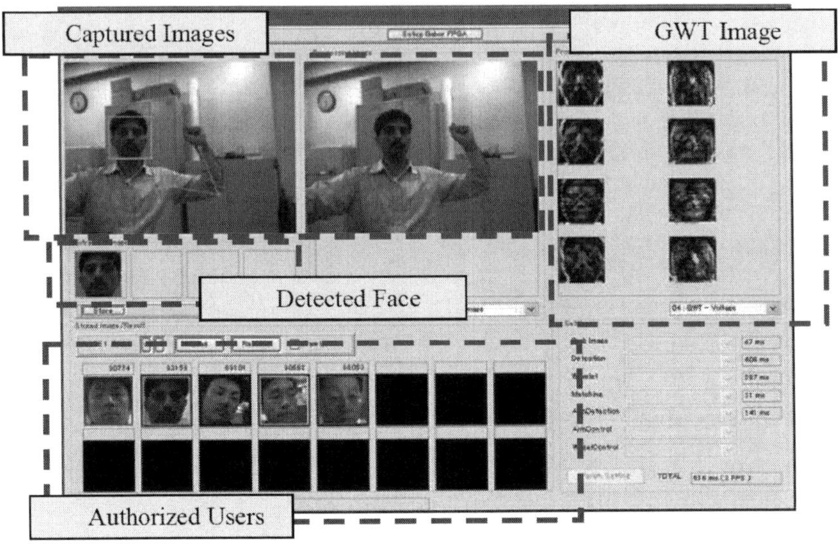

Fig. 2. Screen shot of our system.

4 Implementation

A Screen shot of our system, implementing the above algorithms is shown in Fig. 2. It can display the captured images, detected edges, extracted faces and their GWT coefficients, and faces of the authorized users stored in the database. Position of the detected face and arm are indicated respectively by a rectangle and two lines drawn on the input image. Line color is the same as that of the rectangle around the matching face in the database.

Acknowledgement

This work was partly supported by fund from MEXT, Japan, via a 21st Century COE program (center #J19) granted to Kyushu Institute of Technology.

References

1. `http://sourceforge.net/projects/opencvlibrary`
2. Lades, M.: Distortion Invariant Object Recognition in the Dynamic Link Architecture. IEEE Trans. Computer 42(3), 300–311 (1993)
3. Nagano, S., Morie, T., Nakano, T., Nakamura, K.: A Real-time Image Processing System Using a Gabor-filtering LSI Realizing Primary Visual Cortex Functions. In: Joint 3rd Int. Conf. Soft Computing & Intelligent Systems and 7th Int. Symp. Advanced Intelligent Systems (SCIS & ISIS 2006), Tokyo, Japan, pp. 516–519 (2006)

4. Nakano, T., Morie, T.: A Digital LSI Architecture of Elastic Graph Matching and its FPGA Implementation. In: Proc. Int. Joint Conf. Neural Networks (IJCNN 2005), Montreal, Canada, pp. 689–694 (2005)
5. Shimizu, M., Yoshizuka, T., Miyamoto, H.: A Mobile Robot Controlled by Gesture with Stereo Vision. In: Proc. Brain IT, Kitakyushu, Japan, p. 49 (2006)
6. Viola, P., Jones, M.: Rapid Object Detection using a Boosted Cascade of Simple Features. In: Proc. IEEE Conf. Computer Vision & Pattern Recognition (CVPR), Hawaii, USA, vol. 1, pp. I-511- I-518 (2001)

Shadow Elimination Mimicking the Human Visual System

Takuji Kamada, Akitoshi Hanazawa, and Takashi Morie

Department of Brain Science and Engineering, Graduate School of Life Science and Systems Engineering, Kyushu Institute of Technology, 2-4 Hibikino, Wakamatsu, Kitakyushu 808-0196, Japan
hanazawa@brain.kyutech.ac.jp

Abstract. Shadow elimination is required especially when processing images captured outdoors for object recognition. Whereas room lights make diffuse shadows, the sunlight makes strong shadows leading to misrecognition of objects or some patterns such as humans or traffic signs. To remove this harmful effect of shadows on image processing, we developed a simple algorithm mimicking a human psychological characteristic about shadow detection.

1 Introduction

When recognizing an object based on edge information, shadows behave as noises. Sometimes, a shadow image can be detected as an object by an edge based object recognition system. The human visual system reduces this possibility of misunderstanding by recognizing shadows. How our human vision detects shadows? Several characteristics of shadow are known to be used for shadow recognition, and one of which is a correlation between luminance change and color change. At the border of an object, both luminance and color change at the same place. On the other hand, at the border of a shadow, only luminance changes and color does not change. Thus when we see an image in which a luminance pattern and a color pattern coincide each other, the luminance pattern can be seen as that caused by change in surface reflectance. When the patterns of luminance and color are not aligned, the luminance pattern can be seen as a shadow.

To calculate the correlation between changes in luminance and color, we implemented the representation of color and luminance information employed in human visual system. An original image is decomposed into red-green modulation image, blue-yellow modulation image, and luminance modulation image. Shadows can be seen in the luminance modulation image but not in the color modulation images. Thus the edges detected in the luminance image do not appear in the color images if they are the borders of the shadows. Only the borders of the objects appear as edges.

A. Hanazawa et al. (Eds.): Brain-Inspired Info. Technology, SCI 266, pp. 147–151.
springerlink.com

2 Retinal Model

Human visual system decomposes the information of color and luminance at the level of retinal ganglion cells. As shown in Fig.1, Ganglion cells receive information from three types of cones, which are photosensitive cells working as the front end of the visual system. They respectively are sensitive to short, middle, and long wave length illumination. At the level of cones, color and luminance information are intermingled. For example, red light activates cones sensitive to long wavelength. However, luminance change of the red light decreases the activity as well as color change from red to green does. At the level of the ganglion cells, color-opponent parvocellular cells behave as color-luminance decomposer. For diffuse light covering entire receptive field, red-on green-off cells responds to color shift from green to red but does not responds to luminance increase or decrease at any color, which means that this type of cells extract pure color information from the combination of response amplitudes of L and M cones. On the other hand, when an edge traverses across the receptive field, color opponent cells responds to a luminance difference at the edge, meaning that they behave as spatial band-pass filter for luminance distribution.

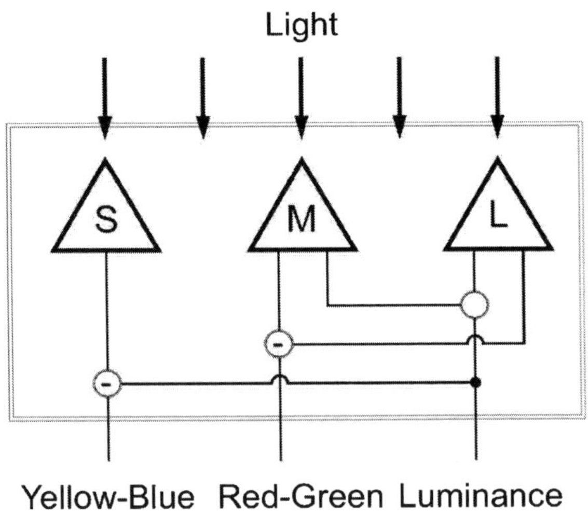

Fig. 1. Color and luminance decomposition in the retina.

3 Shadow Cue

We human discriminate shadows from changes of surface reflectance. One of important cues for the discrimination is correlation between color change and luminance change. As shown in Fig.2, at the borders of shadows, surface reflectance does not change, thus only luminance changes but not color. On the other hand, at the borders of objects and backgrounds, surface reflectance changes

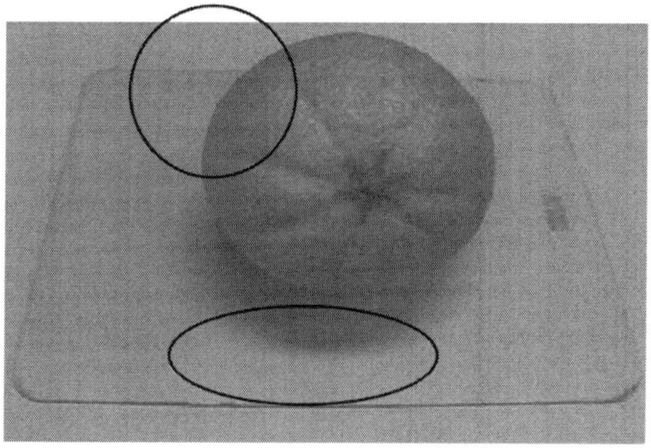

Fig. 2. Whether color and luminance change at the same place is one of strong shadow cues.

resulting in that both luminance and color change. A psychophysical experiment showed that this cue is employed in the human visual system, which strongly influences the perception of three-dimensional shape from shading [1].

4 Image Processing

To represent color information and luminance information independently like the human visual system, RGB color space was transformed into cone color space through CIE color space. Correspondence between RGB color space and CIE color space was calibrated by measuring computer display by a colorimeter. Transformation of color space from CIE color space to cone color space was performed by the method of Lucassen & Walraven[2]. Further, cone color space (Fig.3a) was transformed into perceptual color space (Fig.3b) by projecting colors on the isoluminant plane at 20cd/m^2.

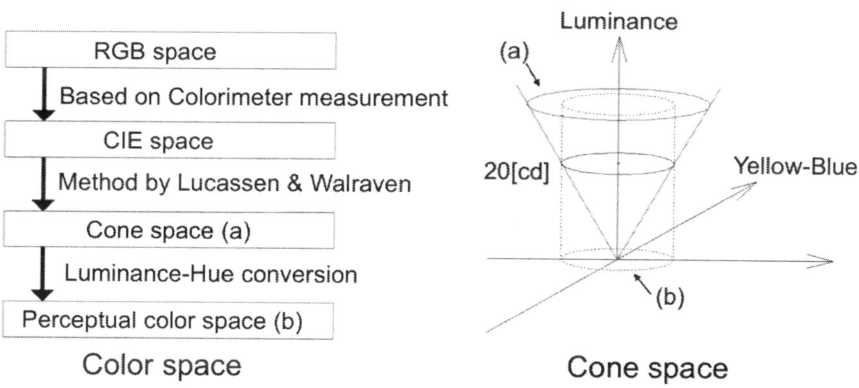

Fig. 3. RGB image values are transformed to represent color and luminance independently.

5 Experimental Results

Images were captured by a video camera connected to a computer (Fig.4 left). Clock frequency of the computer was 3.4 GHz. Frame rate of the camera was 30 fps and image capture speed of the computer was more than the rate. Image size was 100 x 100 pixels. In a case, a shadow was made on a table tennis racket by a hand by interrupting the light (Fig.4 right).

Luminance image, y/b chromatic image, and r/g chromatic image were produced from an original color image (Fig.5). Luminance image is almost the

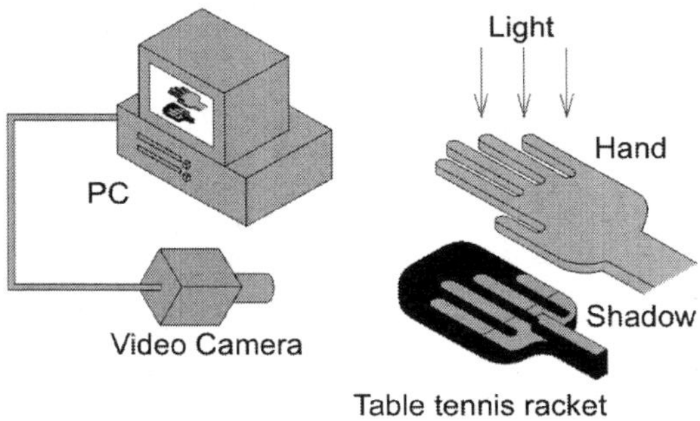

Fig. 4. Experimental setup.

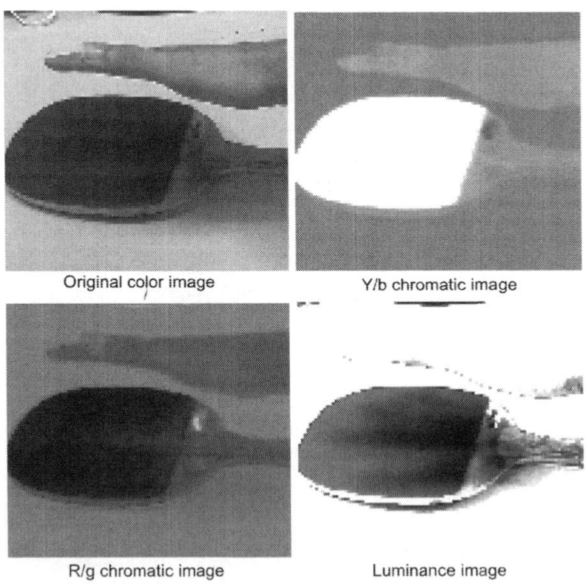

Fig. 5. Shadows in luminance image and chromatic images.

same as ordinary gray scale image produced from color image. Chromatic images, r/g and y/b, represent modulation amplitude from white point along r/g and y/b chromatic axis by gray scale. For luminance image, a shadow that has a shape similar to the hand can be seen like as original color image. For both chromatic images, shadow is eliminated because these images show only changes in color but not luminance.

6 Conclusions

We proposed a shadow elimination algorithm mimicking the human visual system. Because our method is simple, shadows can be eliminated in real time by using a high-performance PC, and can be used for a preprocessing of image recognition algorithms. By using correlation between the luminance and chromatic images, it may by possible to recover original color and luminance from those covered by shadows.

Acknowledgement

This work was supported by a COE program (center #J19) granted to Kyushu Institute of Technology by MEXT of Japan.

References

1. Kingdom, F.A.: Color brings relief to human vision. Nat. Neurosci. 6, 641–644 (2003)
2. Lucassen, M.P., Walraven, J.: Quantifying color constancy: evidence for nonlinear processing of cone-specific contrast. Vision Res. 33, 739–757 (1993)

An FPGA-Based Collision Warning System Using Moving-Object Detection Inspired by Neuronal Propagation in the Hippocampus

Haichao Liang, Youhei Suzuki, Takashi Morie*, Kaziki Nakada, Tsutomu Miki, and Hatsuo Hayashi

Department of Brain Science and Engineering, Graduate School of Life Science and Systems Engineering, Kyushu Institute of Technology, 2-4 Hibikino, Wakamatsu, Kitakyushu 808-0196, Japan
morie@edu.brain.kyutech.ac.jp

Abstract. In this paper, we propose an FPGA-based collision warning system for advanced automobile driver assistance systems or autonomous moving robots. The system consists of three function blocks: edge detection, moving-object detection and danger evaluation and collision warning. In the moving-object detection, the system uses a moving-object detection algorithm inspired by neuronal propagation in the hippocampus, which can run in high speed and low calculation cost. We have applied the system in a robot. It can detect moving objects with a speed range of 3-47cm/s with a sampling period of 33ms for an input image of 320×240 pixels, and can output a warning against dangerous regions in the input image.

1 Introduction

Moving-object detection techniques are essential to collision warning systems for advanced automobile driver assistance systems or autonomous moving robots. So far, optical flow models are mainly used for that purpose. These models used stereo camera systems to calculate the distance and velocity of moving objects by the parallax [5, 6], or used single camera images to calculate these distance and velocity with some complicated mathematical models [1, 7]. Such high calculation cost made collision warning systems slow or expensive. Therefore, simple and high speed algorithms for moving-object detection are needed.

We have already proposed such an algorithm using single-camera images inspired by neuronal propagation in the hippocampus [4, 8], and have already implemented the algorithm in an FPGA [2]. In this paper, we propose a real-time collision warning system based on the FPGA implementation of our moving-object detection algorithm.

---

* Corresponding author.

A. Hanazawa et al. (Eds.): Brain-Inspired Info. Technology, SCI 266, pp. 153–158.
springerlink.com

2 FPGA-Based Collision Warning System

Our system consists of three function blocks: edge detection, moving-object detection, and danger evaluation and collision warning, as shown in Fig. 1. In edge detection, we use some digital image processing to reduce noise in input images and detect principal edges as much as possible. In danger evaluation and collision warning, we use fuzzy inference to analyze the moving-object detection results and evaluate the degree of danger.

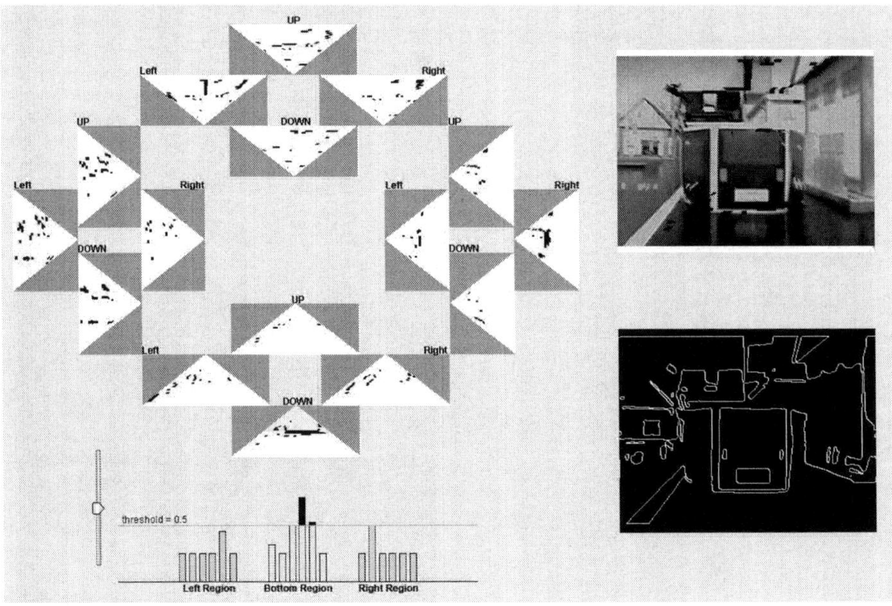

Fig. 1. System operation results using actual images.

Compared with conventional approaches, our system has an advantage of high speed operation and low calculation cost because it performs neither image matching processing nor actual velocity calculation. Our system only treats object edges instead of objects themselves and evaluates the degree of danger of predefined regions. Even if an occlusion occurs, it hardly affects the evaluation results. Our system detects the speed of moving objects in each predefined direction instead of calculating the actual velocity, which also reduces the calculation cost. Therefore, we can develop a faster and more compact collision warning system.

2.1 Edge Detection

In this system, we use three kinds of filters to detect principal edges: Mean filter, Gaussian filter, and Difference filter. Firstly, we use a mean filter to reduce impulse noise, and then use a Gaussian filter to reduce Gaussian noise. Both of them will also smooth the input images, and small regions disappear. Finally, we use a difference filter to detect edges. An edge detection result is shown in Fig. 1.

2.2 Moving Object Detection

The hippocampus is known as a brain area related with memory and spatial navigation in humans and animals. The hippocampal formation consists of three principal regions: the dentate gyrus (DG), CA3 and CA1. We proposed a moving-object detection algorithm inspired by neuronal propagation underlying sequence coding in a model of the hippocampus [8].

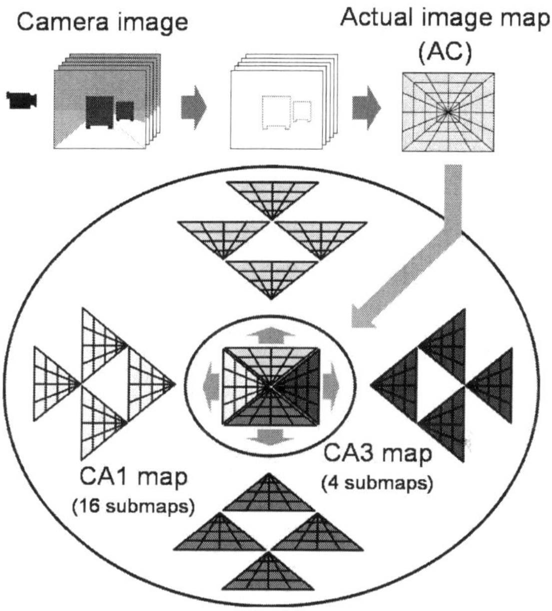

Fig. 2. Moving object detection model inspired by neuronal propagation in hippocampus.

The proposed algorithm employs three kinds of maps: actual image (AC) map, CA3 map and CA1 map. They are divided into pieces based on radial coordinates and concentric rectangles as shown in Fig. 2. The CA3 map is divided into four submaps, and each CA3 submap is related to four corresponding CA1 submaps. We call a piece of them a unit, because it usually includes tens of pixels. The edge information of the input image is mapped onto the AC map. The AC map then generates unit-firing information which determines whether a CA3 unit fires or not according to the edge location in the corresponding AC unit. Each CA3 submap receives the unit-firing information and sends signals to four corresponding CA1 submaps that are specified to detect and represent objects moving in the direction of up, down, right and left by a pattern of activated units, respectively. Decaying of the value of each CA1 unit is used, instead of neuronal propagation, to measure how long it takes for each edge of moving objects to pass through a unit to the neighboring one, which is hereafter called passing time.

In the digital VLSI architecture design, we use a digital counter for the decaying signal generator to measure the passing time. In addition, the counter is shared with all units to reduce power dissipation. The detection principle of upward movement using our algorithm is shown in Fig. 3. If a CA3 unit receives a firing signal, it fires and holds the counter value at that time, and then sends the value to the neighboring CA1 units. Movement detection is accomplished by calculating the counter value difference between each CA3 unit and the corresponding CA1 unit in each detection direction. The passing time can be obtained from the counter value difference, which corresponds to the relative velocity because the distance between the neighboring units is fixed. We can detect the relative speed of multiple moving objects in each predefined direction in high speed by using this simple algorithm. We have implemented this algorithm in

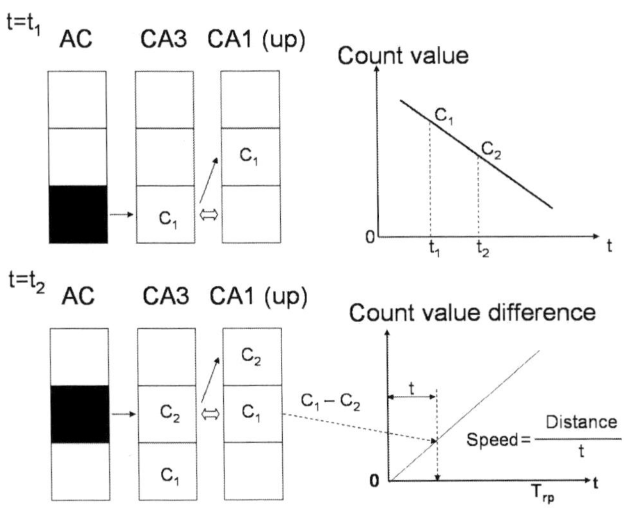

Fig. 3. Detection of upward movement using our algorithm.

Table 1. Specifications of moving object detection part.

Image size	320 × 240 pixels (Width × Height)
Sampling period	33 ms
Number of units	32 × 32 /submap
CA1 decaying steps	64 steps (6-bit counter)
CA3 Refractory period	2,112 ms (33 ms × 64)
Passing time (unit to unit)	33 – 2,112 ms
Speed range of detection	3 – 47 cm/s

an FPGA [2], and the specification of the moving object detection part using an FPGA is shown in Table 1.

2.3 Danger Evaluation and Collision Warning

In the danger evaluation, we use fuzzy inference to analyze the moving-object detection results and evaluate the degree of danger. We additionally divide each CA1 submap into two parts for that purpose, as shown in Fig. 4A. The height or width of a CA1 submap is divided into half respectively. We call black regions region-1 and white regions region-2.

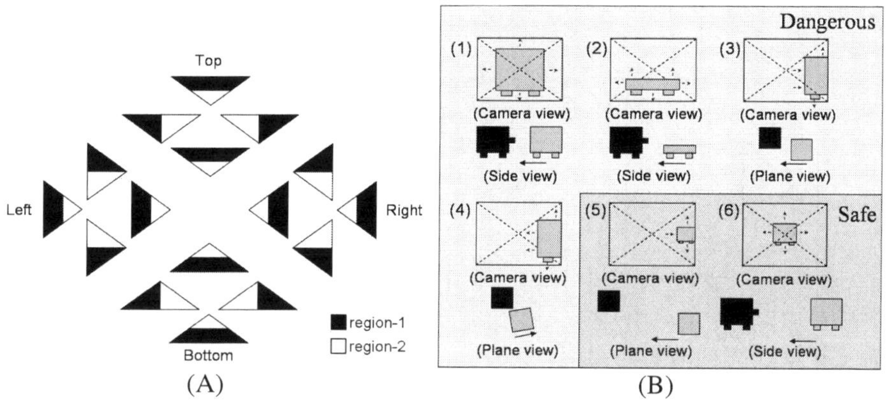

Fig. 4. (A) CA1 submap segmentation for danger evaluation, (B) Car-to-car moving patterns for the collision warning system.

We explain the danger evaluation processing by using car-to-car moving patterns shown in Fig. 4B. For front impact, as shown in Fig. 4B (1) and (2), downward moving edges are detected in the bottom-region-1, and leftward and rightward moving edges are detected in the left-region-1 and the right-region-1, respectively. For front-to-side impact, as shown in Fig. 4B (3) and (4), downward moving edges are also detected in the bottom-region-1. In safe cases shown in Fig. 4B (5) and (6), downward moving edges are not detected in the bottom-region-1. Thus, we can define some if-then rules for evaluating the degree of collision danger using the detected moving edges according to their average speed, acceleration and the features described above. For more detail, please refer to our corresponding proceeding paper [3].

An example of fuzzy inference results is shown in Fig. 1. The fuzzy inference result for each CA1 submap is expressed in the range from 0 to 1, and we set here the threshold value for danger evaluation as 0.5. In other words, if the fuzzy inference result is larger than 0.5, the system will output a warning using the predefined red color as shown in Fig. 1. We verified the system operation using an autonomous moving robot, and the system successfully output a warning against the approaching objects with red regions as shown in Fig. 1.

3 Conclusion

We have proposed an FPGA-based collision warning system for advanced automobile driver assistance systems or autonomous moving robots, and have applied our system to an autonomous moving robot to verify the system operation for real-world complicated moving images. We successfully verified the collision warning operation for some simple car-to-car moving patterns. In the future work, we will improve the fuzzy inference for danger evaluation, and verify the system operation for more complicated car-to-car moving patterns.

Acknowledgment

This work was partly supported by fund from MEXT, Japan, via a 21st Century COE program (center #J19) granted to Kyushu Institute of Technology.

References

1. Brailon, C., Pradalier, C., Crowley, J.L., Laugier, C.: Real-time moving obstacle detection using optical flow models. In: IEEE Intelligent Vehicles Symposium, pp. 466–471 (2006)
2. Liang, H., Morie, T., Nakayama, H., et al.: An FPGA-based real-time moving-object detection system based on neuronal propagation in the hippocampus. In: RISP NCSP 2007, pp. 509–512 (2007)
3. Liang, H., Morie, T., Suzuki, Y., et al.: An FPGA-based Collision Warning System Using Hybrid Approach. In: 7th Int. Conf. on Hybrid Intelligent Systems, pp. 30–35 (2007)
4. Liang, H., Nakayama, H., Nakada, K., et al.: Digital VLSI implementation of a moving object detection algorithm based on neuronal propagation in the hippocampus. In: ISPACS 2006, pp. 614–617 (2006)
5. Nedevsch, S., Danescu, R., Frentiu, D., et al.: High accuracy stereo vision system for far distance obstacle detection. In: IEEE Intelligent Vehicles Symposium, pp. 292–297 (2004)
6. Shima, N., Babasaki, M., Akidzuki, Y., et al.: Fusion sensor for driving assistance system. Fujitsu Ten Technical Journal (17), 35–44 (2001)
7. Yokohama, M., Poggio, T.: A contour-based moving object detection and tracking. In: Proceeding of Second Joint IEEE International Workshop on Visual Surveilance and Performance Evaluation of Tracking and Surveilance (VS-PETS), pp. 271–276 (2005)
8. Yoshida, M., Hayashi, H.: Organization of cell assemblies that code temporal sequences in a hippocampal CA3-CA1 model. In: IJCNN 2004, pp. 495–500 (2004)

A Chemical Sensor Array Inspired by Mouse Taste Buds

Jun Igarashi, Katsumi Tateno, Kazuki Nakada, Tsutomu Miki,
Yoshitaka Ohtubo, and Kiyonori Yoshii

Department of Brain Science and Engineering, Graduate School of Life Science and Systems Engineering, Kyushu Institute of Technology, 2-4 Hibikino, Wakamatsu, Kitakyushu 808-0196, Japan
{igarashi,tateno,nakada,miki,otsubo}@brain.kyutech.ac.jp,
yoshii@brain.kyutech.ac.jp

Abstract. We made a chemical sensor array that detected the concentration of alcohol gas as the degree of synchronization of neural activity. The chemical sensor array consisted of three parts: the chemical sensor, the random pulse generator, and the stochastic synchronizer. The chemical sensor part was commercially available alcohol sensors. The random pulse generator and the stochastic synchronizer parts were implemented in a computer. Outputs of alcohol sensors charged integrate-and-fire cells in the random pulse generator part and caused these cells to fire spikes. These spikes equally converged on a pair of bursting cells in the stochastic synchronizer part and led to the synchronization of their bursting activity. The alcohol gas concentration monotonically increased the frequency of the random pulses and then increased the degree of burst synchronization on the two burst cells. As a result, the concentration of alcohol gas was detected as the degree of burst synchronization.

1 Introduction

Each taste bud, a taste-sensing organ consists of several tens of cells, which are classified into four types of cells (Type I to Type IV). The taste nerve has chemical synapses on only Type III cells. It means that Type III cell is the output cell of a taste bud. On the other hand, Type II cells have taste receptors for bitter, sweet, or umami. As Type II cells do not have synapses of the taste nerve, communications between Type II cells and Type III cells are necessary for taste signal processing. Ohtubo and Yoshii have shown possible signal transmissions using chemical transmitters or gap junctions between TBCs [2]. That result suggests that the network in a single taste bud contributes to process taste information on tongues.

We have proposed a computational model of a chemical sensor array which consisted of integrate-and-fire (LIAF) cells and bursting cells. The LIAF cell and the bursting cell corresponded to the Type II cell and the Type III cell respectively. We assumed that chemical substances charged the LIAF cells. The LIAF cells

A. Hanazawa et al. (Eds.): Brain-Inspired Info. Technology, SCI 266, pp. 159–164.
springerlink.com

showed irregular spiking in response to the chemical stimuli. Spikes of the LIAF cells equally converged on a pair of non-identical uncoupled bursting cells. Common random pulses stochastically entrained bursts in the two bursting cells [3]. The degree of burst synchronization depended on the frequency of the spike intervals in the LIAF cells and the shape of their histogram. When the histogram of the random pulses was an exponential-like distribution and its frequency was high enough, the burst synchronization in the two bursting cells was facilitated.

The above chemical sensor array needs receptor membranes which charge the membrane potential of the LIAF cells. In the present study, the chemical sensor system was demonstrated with commercially available alcohol sensors. The alcohol gas sensors had non-uniform responses to the alcohol stimuli. The present results showed that the non-uniformed sensitivity of the sensors was necessary for synchronous burst responses of the bursting cells. Further, we redefined the components of the chemical sensor array as the random pulse generator and the stochastic synchronizer parts(Fig. 1a). The functions on each part were generalized in the present study.

2 Methods

One of possible implementations of our sensor array is a computational model with actual chemical sensors (Fig. 1b). In the present study, commercially available alcohol gas sensor, TGS2600 (FIGARO engineering inc.), was used in the chemical sensor. The present chemical sensor array consisted of 32 alcohol sensor circuits (chemical sensor), 32 leaky integrated-and-fire (LIAF) cells (random pulse generator) and a pair of non-identical uncoupled bursting cells (stochastic synchronizer) (Fig. 1b). No connection existed among the LIAF cells.

The alcohol sensors were set in a plastic bottle. The alcohol sensor has the internal resistance of 10 kΩ - 90 kΩ in air. The internal resistance decreased in alcohol gas. A change in the internal resistance of the alcohol sensors was converted to the voltage by op-amp. The output voltage of the alcohol sensor circuits was normalized by the equation below:

$$s = \frac{\max_{\mathrm{sim}} - \min_{\mathrm{sim}}}{\max_{\mathrm{sensor}} - \min_{\mathrm{sensor}}} \cdot (V_{\mathrm{s}} - \min_{\mathrm{sensor}}) + \min_{\mathrm{sim}}$$

where $\max_{\mathrm{sim}} = 0.014$, $\min_{\mathrm{sim}} = 0.007$, $\max_{\mathrm{sensor}} = 4.0$, $\min_{\mathrm{sensor}} = 0.5$.

The output s_i from one of the alcohol sensors charged a LIAF cell. When the membrane potential u reaches or exceeds a threshold, the LIAF cell fires. The LIAF cell is described by the equation below:

$$\frac{du_i}{dt} = -\frac{1}{\tau} u_i + s_i$$

where $\tau = 1000$ ms. The frequency of output pulses monotonically increased with the increase in the alcohol gas concentration. Practically, the chemical sensors received noise during the experiments. For small noise amplitude, an interval distribution of the integrate-and-fire cell is approximately given by a Gaussian when superthreshold stimulation is applied [4]. Those pulses converged on the bursting cells. A total interval distribution from the 32 LIAF cells showed an exponential distribution (CV ~ 1.0).

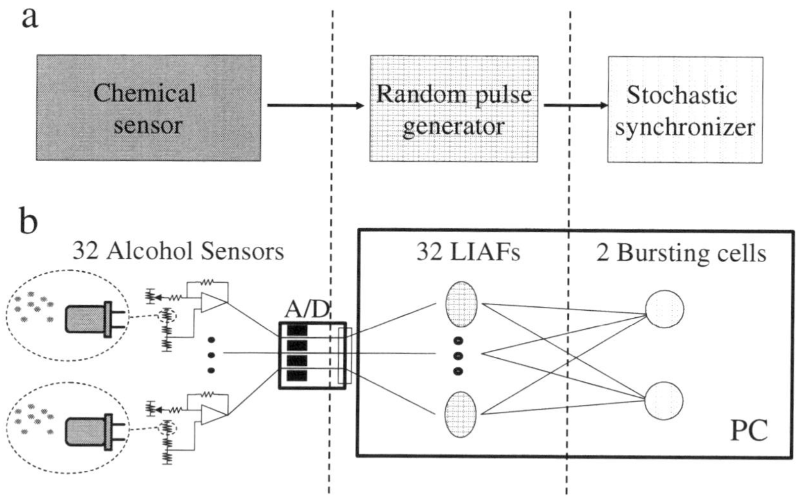

Fig. 1. Implementation of a chemical sensor array with alcohol sensors. a) Diagram of the chemical sensor array. b) 32 alcohol sensors constituted the chemical sensor part. Leaky integrated-and-fire (LIAF) cells constituted the random pulse generator part. Two uncoupled bursting cells constituted the stochastic synchronizer part.

The bursting cell is described by a canonical model proposed by Ermentrout and Kopell [1] and consists of fast (θ) and slow (ψ) subsystems.

$$\frac{d\theta_i}{dt} = (1-\cos\theta_i) + (1+\cos\theta_i)\cdot\cos\psi_i$$

$$\tau\frac{d\psi_i}{dt} = 1-\varepsilon_i\sin\psi_i + w\sum_k \delta_k(t)$$

Pulses from the LIAF cells excited the slow subsystem. Parameter ε_i was 1.20 and 1.22 for the bursting cells #1 and #2 respectively. The connection weight (w) was 0.0944. τ = 200 ms. When $\varepsilon > 1.0$, ψ has a stable node. In the absence of the input, ψ stayed at the stable node and θ showed repetitive oscillations. The random pulses from the LIAF cells induced the saddle-node bifurcation in ψ. When cos $\psi < 0$, the repetitive oscillations of θ were terminated and thus bursts of oscillations appeared. A slight difference of ε caused differences in the interburst intervals. A high rate of the random pulses shortened the interburst intervals. As a result, the interburst intervals of the two bursting cells converged. This led to similar slow dynamics in the two bursting cells. Therefore, a high concentration of alcohol gas caused synchronous responses in the bursting cells.

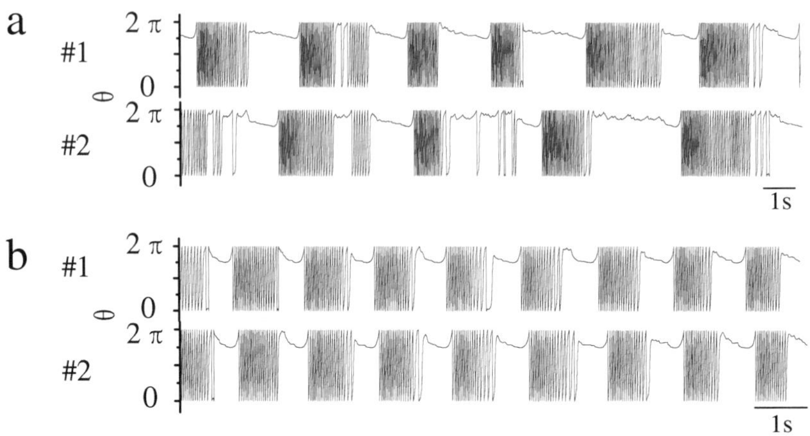

Fig. 2. Synchronization of burst firing. a) 0.14 % of alcohol gas. Burst firings were induced in a low alcohol gas and were not synchronized. b) 2.8 % of alcohol gas. The bursting cells simultaneously showed burst firings.

Before recording, we adjusted outputs of the all alcohol sensors to 0.5V in the air as zero point adjustment. Alcohol gas was injected into the plastic bottle. After 60 sec, the personal computer acquired output voltages of the alcohol gas sensors through an A/D converter. The sampling frequency was 500 Hz. Numerical integrations of the LIAF and the bursting cells also started. Oscillations of the fast subsystem θ on the bursting cells were collected for 300 sec. The experiments repeated 5 times at each concentration of the alcohol gas. Synchronization of bursts was quantified by calculating the synchronization index as $\gamma^2 = \langle\cos(\Delta\phi)\rangle^2 + \langle\sin(\Delta\phi)\rangle^2$, where <> denoted the average over time.

Results in the integrations were displayed in real-time. The specification of the personal computer was as follows: 3.6 GHz Pentium4 CPU, 1GB RAM, and GeForce7600GS for GPU. OS was Windows XP. Graphical user interface was developed by Microsoft Visual C++ with Microsoft Foundation Classes. We used the Euler scheme for solving the differential equations. Step size was 0.02 ms.

3 Results

The alcohol gas injection caused an elevation of output voltages of the alcohol sensors. The output voltages were different in every sensor circuit. The coefficient of variation (CV) of the output voltages was about 0.08. Those sensor outputs caused asynchronous firing in the LIAF cells. Low concentration of the alcohol gas induced firing in few LIAF cells. The mean interval of total spikes of the LIAF cells was 12 ms at 0.14 % of alcohol gas. Bursts in the two bursting cells had different intervals (Fig. 2a). Therefore, bursts were not synchronized.

The increase in the concentration of the alcohol gas shortened the interburst intervals on the LIAF cells. The mean interval of the total output pulses was 3.3 ms at 2.8 % of alcohol gas. That histogram was an exponential distribution (CV = 0.96). The burst intervals shorten due to high frequency inputs of the random pulses. Consequently, synchronization of bursts between the bursting cells was facilitated (Fig. 2b).

The degree of burst synchronization increased with the increase in the alcohol gas concentration (Fig. 3). The synchronization index γ gradually increased and saturated. As a result, the concentration of alcohol gas was encoded to the degree of synchronization of the bursts.

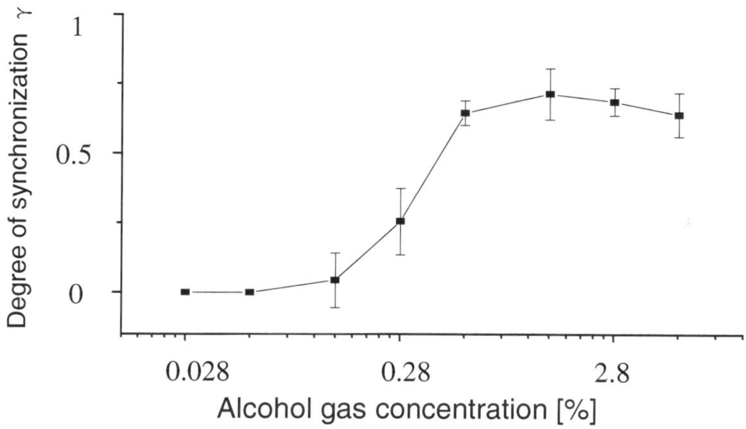

Fig. 3. Conversion of alcohol gas concentration into degree of synchronization

4 Discussion

The present chemical sensor array properly worked with non-uniform chemical sensors. In the previous study [3], when CV of the chemical sensor output was assumed to be 0.2, the stochastic synchronizer part showed burst synchronization. This indicates that the present chemical sensor array was robust enough to replace some of chemical sensors with new ones without compensation, which recalls the physiological turnover of taste bud cells.

We also designed an analog integrated circuit of resonate-and-fire cell as the stochastic synchronizer part for chemical sensor array. Uncoupled resonate-and-fire cells were entrained by a sequence of random pulses with Gamma-distributed interpulse intervals. The detailed description appears later in this book.

Acknowledgement

This work was supported by a COE program (center #J19) granted to Kyushu Institute of Technology by MEXT of Japan.

References

1. Ermentrout, G.B., Kopell, N.: SIAM Journal on Applied Mathematics 46, 233–253 (1986)
2. Ohtubo, Y., Suemitsu, T., Shiobara, S., Matsumoto, T., Kumazawa, T., Yoshii, K.: Journal of Physiology 530, 287–293 (2001)
3. Tateno, K., Yoshii, K., Ohtubo, Y., Miki, T.: International Congress Series 1301, 52–55 (2007)
4. Tuckwell, H.C.: Introduction to theoretical neurobiology. Cambridge University Press, Cambridge (1988)

Emotional Behavior and Expression Based on a Neural Network Model of Amygdala

Satoshi Sonoh*, Shuji Aou, Keiichi Horio, Hakaru Tamukoh, Takanori Koga, Naoki Shimo, and Takeshi Yamakawa

Department of Brain Science and Engineering, Graduate School of Life Science and Systems Engineering, Kyushu Institute of Technology,
2-4 Hibikino, Wakamatsu, Kitakyushu 808-0196, Japan
sonoo-satoshi@edu.brain.kyutech.ac.jp

Abstract. In this research, we propose an emotional expression model inspired by amygdala that governs the processing of emotional information. The proposed model achieves two functions; one is discrimination of sensory inputs, and the other is classical conditioning. We also develop a specific digital hardware of the model and integrate an emotional expression system in a real-world robot.

1 Introduction

By letting intelligent robots generate an emotional expression corresponding to their internal statuses, a man-machine-interaction (MMI) such as an interactive training of robot's movement will be singularly evolved. This is because non-verbal communication based on emotional expression is universal. In order to let robots generate emotional expressions, it is useful to learn from animal's brain. The amygdala in the limbic-system of brain governs emotional information processing [1]. The amygdala has two key functions, one is discrimination of sensory inputs, and the other is classical conditioning between a stimulus that directly induces emotions and the sensory inputs. Consequently, the emotional expression is realized by these two functions.

Although computational models of the amygdala have been proposed to find out emotional learning systems in anatomy [2] and psychology [3], there are no successful applications to engineering and robotics areas. In this paper, we propose an engineering model inspired by the amygdala as a novel emotional expression system for robots. The proposed model emulates two functions, i.e. discrimination of sensory inputs and classical conditioning. In the traditional models, a classification of the sensory inputs must be accomplished in advance. In contrast, the proposed model can adaptively classify the sensory input by using a self-organizing map. This feature is very essential for robots in dynamical environments. We verify effectiveness of the model by a basic software simulation. Finally, we develop a specific digital hardware of the model and integrate an emotional expression system in real-world robot. In experiments with

* Corresponding author.

A. Hanazawa et al. (Eds.): Brain-Inspired Info. Technology, SCI 266, pp. 165–170.
springerlink.com

the robot, an emotional interaction between the human and the robots can be achieved. Furthermore, through the interaction, the robot works out to generate various emotions for conditioned sensory inputs.

2 Emotional-Expression Model of Amygdala: EMA

Emotional-expression Model of Amygdala (EMA) that we propose as an artificial neural network of the amygdala can provide the emotional expression capability for robots. Fig. 1 shows the architecture of EMA based on anatomical findings of the amygdala. EMA consists of three layers, sensory input layer, LA layer, and CE layer. The sensory input layer includes several input units modeling entry areas to the amygdala. The LA layer has a number of functional units which are arranged in one or two dimensional array and forms a feature map of the sensory inputs. The CE layer is the output layer of EMA. The output is emotional values which are defined as strength of several emotions (e.g., fear, pleasure, surprise and so on). EMA contains two computational processes which are the stimulus recognition process and the conditioning process. We will detail these processes in the following section.

2.1 Stimulus Recognition Process

In the stimulus recognition process, a large amount of sensory inputs is integrated and discriminated by using an adaptive-tuned self-organizing map. Let $\boldsymbol{x}$(t) be a sensor y input vector at time step t and $\boldsymbol{w}_i(t)$ a reference vector of i-th unit on the LA layer. The best matching unit (BMU) for the sensory input is selected as follows:

$$BMU(t) = \arg\min_i(\|\boldsymbol{x}(t) - \boldsymbol{w}_i\|^2) = \arg\min_i(error_i(t)).$$

BMU represents the recognized sensory input and the reference vectors are updated by competitive learning.

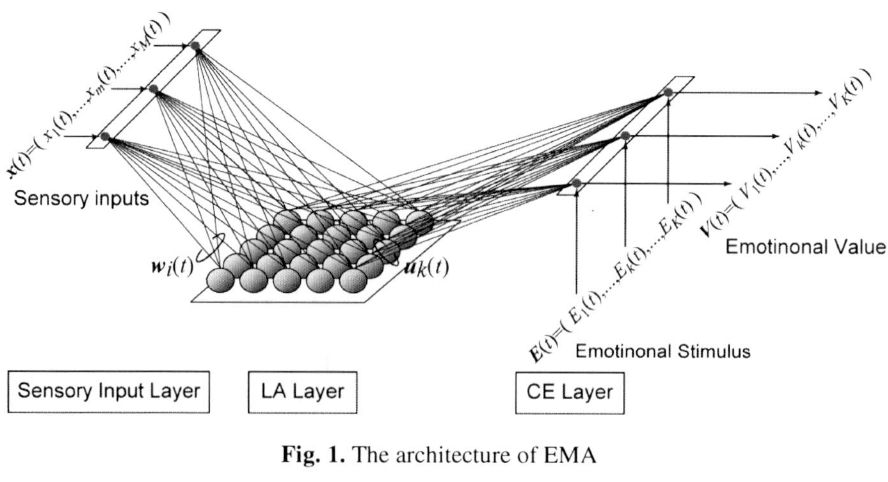

Fig. 1. The architecture of EMA

$$\boldsymbol{w}_i(t+1) = \boldsymbol{w}_i(t) + \alpha(t) h_{i,BMU}(t)\big(\boldsymbol{x}(t) - \boldsymbol{w}_i(t)\big),$$

$$h_{i,BMU}(t) = \exp\left(\frac{D(i, BMU)^2}{2\sigma(t)^2}\right).$$

Here, $D(i, BMU)$ is the distance function between i-th unit and BMU on the LA layer. $\alpha(t)$ and $\sigma(t)$ are the learning rate and the neighboring width, respectively. Although these parameters usually decrease according to a progress of the time step, these parameters in EMA should be adaptively changed, because we assume applications to robots in dynamical environments. Furthermore, in order to facilitate an implementation of EMA, a simple and convenient approach to the adaptive change is desirable. We adopt a parameter-less method [4] and these parameters are determined according to a relative error as follows:

$$\alpha(t) = \frac{error_i(t)}{error_{\max}},$$

$$\sigma(t) = \max(\alpha(t) * \sigma(t)_{max}, \sigma(t)_{min}).$$

Here, $error_{\max}$ is the maximum error until time step t. $\sigma(t)_{max}$ and $\sigma(t)_{min}$ are the maximum and minimum neighboring widths, respectively. These equations mean that the learning rate and the neighboring width increase where the relative error is large, but decrease in small relative error. Thus, the sensory input is successfully recognized by the adaptive-tuned feature map.

2.2 Conditioning Process

In the conditioning process, relationships between the sensory inputs and an emotional stimulus are acquired by classical conditioning. The emotional stimulus, given from the MMI or the environment, inherently induces a specific emotion. In the Pavlov's experiments, a food fed to the dog corresponds to the emotional stimulus, and a sound of a bell does to the sensory inputs. In EMA, the strength of an emotion induced by the sensory inputs that intrinsically independent to emotions is represented as an emotional value $V(t)$:

$$V_k(t) = \sum_i u_{k,i}(t) \cdot act_i(t),$$

$$act_i(t) = \frac{h_{i,BMU}(t)}{\sum_j h_{j,BMU}(t)}.$$

Here, suffix k is the index of the emotion and $u_{k,i}$ is an associate weight between i-th unit and k-th emotion. The conditioning by the emotional stimulus $E_k(t)$ is represented as follows:

$$u_{k,i}(t+1) = u_{k,i}(t) + \delta \cdot act_i(t) \cdot \big(E_k(t) - V_k(t)\big).$$

Here, is the conditioning rate. The emotional expression in EMA is realized by the conditioning process, because the sensory inputs are dynamically associated to the emotions.

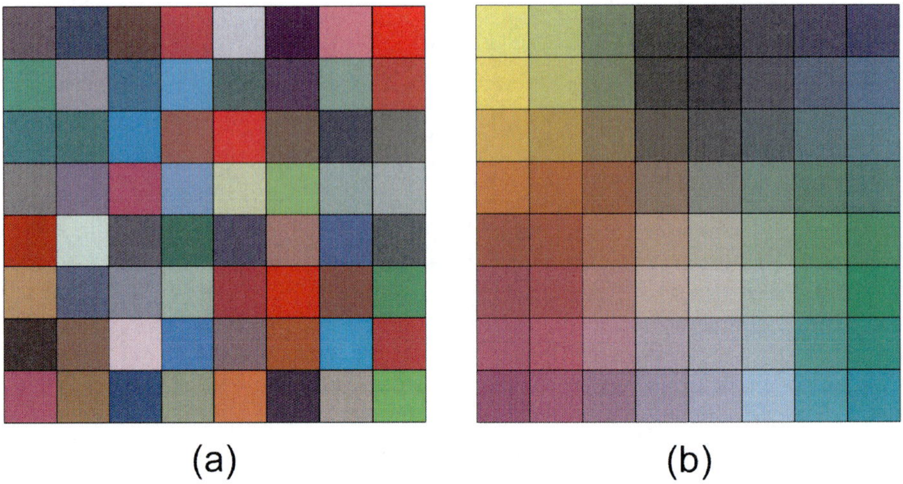

Fig. 2. The feature map of the sensory inputs. A color patch represents a reference vector. (a) Initial state. (b) After 1000 random presentations.

2.3 Computational Simulation

The effectiveness of EMA is verified in a computational simulation. In the simulation, color signals are used as sensory inputs $\boldsymbol{x} = \{x_r, x_g, x_b\}$, and "hitting" and "stroking" as emotional stimuli $\boldsymbol{E} = \{E_h, E_s\}$ correspond to "fear" and "pleasure" emotions. The discrimination and conditioning of EMA are confirmed by using these inputs.

Fig. 2 shows the feature map represented by the reference vectors of 64 units on the LA layer. Fig. 2(a) is randomly initialized feature map and Fig. 2(b) is after 1000 random presentations of sensory inputs. The feature map clearly preserves the topology of the sensory inputs. The topology preservation is important for robust recognition of the sensory inputs in noisy real-environments.

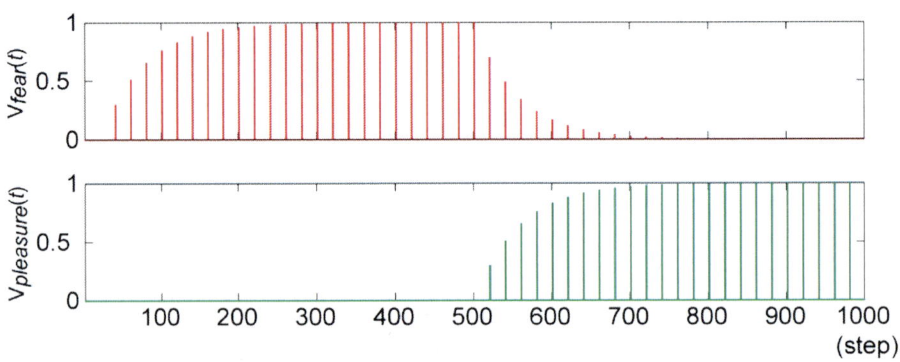

Fig. 3. The emotional value in the basic conditioning experiments.

Fig. 3 shows a result of basic conditioning experiments, "acquisition" and "extinction". The acquisition is a phenomenon that the emotional value enlarges due to simultaneous presentations of the specific sensory inputs and the emotional stimulus. On the other hand, the extinction is a phenomenon that the decrease of the emotional value results from the presentations of the sensory inputs excluding the emotional stimulus. In the experiments, the acquisition for the fear emotion is executed in the first-half steps. Continuously, the extinction for the fear emotion and the acquisition for pleasure emotion are executed in the last-half steps. As a result, EMA can successfully learn each emotional value.

3 Implementation of EMA in Real-World Robot "WITH"

We implement EMA in a real-world robot "WITH". In order to realize a real-time control of the robot, we develop a specific digital hardware of EMA (EMAHW) by using FPGA (Xilinx Spartan-3E, sc3s1600E). The EMAHW has a massively parallel architecture for the number of LA unit, and can process calculations of 81 units at clock frequency of up to 35MHz. The calculation speed is twenty-times faster than one of a portable PC (Intel 1.1GHz, 512M memory).

And also, we develop emotional communication devices for non-verbal MMI. As an input device for the emotional stimulus from the human to the robot, there is a touch sensory array which is able to discriminate either "hitting" or "stroking" on top of the robot (See Fig.4). On the other hand, output devices are a tail-like device and an ear-like device. Emotional reactions from the robot to the human are achieved by movements of these devices according to the emotional values EMAHW generates. The human can understand robot's emotions intuitively, because the movements of ear and tail just imitate behaviors of dogs which are familiar pets.

We demonstrate an emotional interaction training using the robot integrated with the emotional expression system. In the training, some colored-markers as

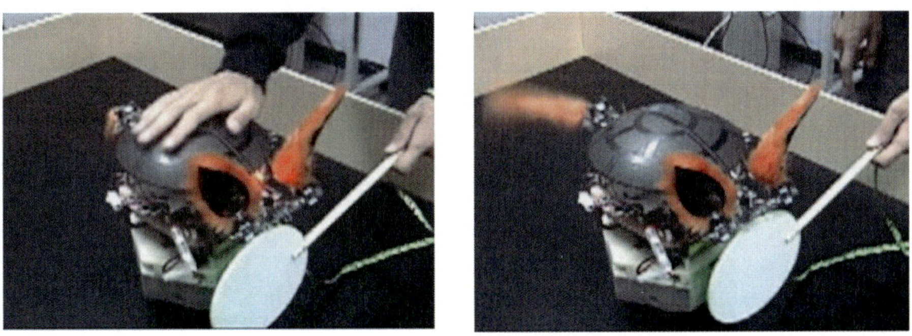

(a) (b)

Fig. 4. A demonstration by using real-world robot with EMA. (a)A scene of conditioning. (b)A scene that the robot expresses its emotions.

the sensory inputs and the emotional stimuli are directly given by hands of a trainer. Fig. 4(a) shows a scene that the trainer makes the robot learn to associate a marker with "stroking". After the conditioning, the robot expresses pleasure emotion by tail wagging when the robot only looks at the marker. This demonstration is very simple, but the robot realizes more emotional and interesting interactions by using combinations of emotions and various behaviors.

4 Conclusions

In this research, from an engineering point of view, we proposed the emotional expression model inspired by the amygdala. We also implemented the model into a digital hardware in order to let the model be easily applicable to various purposes such as robot controllers. As an application example, we realized an emotional human-robot-interaction using real-world robot with proposed emotional expression system.

Acknowledgement

This work was supported by a COE program (center #J19) granted to Kyushu Institute of Technology by MEXT of Japan. We express our sincere appreciations to H. Isogai, A. Hanazawa, H. Nakagawa, T. Toyoshima and M. Mizuno (Professors of KIT LSSE), Prof. N. Kasabov and Dr. S. Pang (KEDRI, AUT) for their enthusiastic discussion about this research.

References

1. Aggleton, J.P.: The amygdala: A Functional Analysis. Oxford University Press, Oxford (2000)
2. Armony, J.L., Servan-Schreiber, D., Cohen, J.D., LeDoux, J.E.: An anatomically constrained neural network model of fear conditioning. Behav. Neurosci. 109, 246–257 (1995)
3. Moren, J., Balcenius, C.: A Computational Model of Emotional Learning in the Amygdala. From Animal to Animats 6, 383–391 (2000)
4. Berglund, E., Sitte, J.: The parameterless self-organizing map algorithm. IEEE Trans. on Neural Netw. 17(2), 305–316 (2006)

Effective and Adaptive Learning Based on Diversive/Specific Curiosity

Naoki Shimo[1,*], Shaoning Pang[2], Keiichi Horio[1], Nikola Kasabov[2], Hakaru Tamukoh[1], Takanori Koga[1], Satoshi Sonoh[1], Hirohisa Isogai[1], and Takeshi Yamakawa[1]

[1] Department of Brain Science and Engineering, Graduate School of Life Science and Systems Engineering, Kyushu Institute of Technology, 2-4 Hibikino, Wakamatsu, Kitakyushu 808-0196, Japan
shimo-naoki@edu.brain.kyutech.ac.jp

[2] Knowledge Engineering and Discovery Research Institute, Auckland University of Technology, Auckland 1020, New Zealand
spang@aut.ac.nz, nkasabov@aut.ac.nz

Abstract. In this paper, an effective and adaptive learning model, in which a concept of curiosity is used, is proposed. The key idea of the proposed method is to introduce the diversive curiosity in addition to specific curiosity. Furthermore we employ the concept of threshold, which control timing of switching two curiosity modes. By employing two curiosity and making a proper selection of them, a learning ability and adaptability is improved. The effectiveness of the proposed method is verified by some simulations.

1 Introduction

Let us assume that an agent (learner) should learn some problems whose difficulty levels are different. Some problems are easy, and some problems are difficult. Initially, the agent tries to learn various problems. To learn some easy problems in the initial stages is sometimes useful for learning a difficult problem, when the problems are mutually relevant. In that case, it is not effective that the agent tries learning the difficult problem at the beginning. If the agent firstly faces a difficult problem, he/she should immediately give up it and try looking for other problems. Human makes proper selections of the problems. In other words, human feels bored learning the problem which is already learned, and human gives up the problem which is too difficult to learn [1,2,3]. In psychological review, it is supposed that the switching point of the search mode of initial stage and the one of the latter stage is important. However, it is not clarified. Thus, we think it important to investigate the switching point. In addition, we are interested also in the implementation of each search mode in artificial learning algorithms.

* Corresponding author.

A. Hanazawa et al. (Eds.): Brain-Inspired Info. Technology, SCI 266, pp. 171–175.
springerlink.com

2 Curiosity

A concept of a curiosity may help to control timing of switching the problems that the agent faces. Curiosity is an intrinsic motivation, and the agent who is hungry for information makes an action to get some information voluntarily. Curiosity is divided to two types, diversive and specific curiosities [1,2,3]. The former is the curiosity which faces to not particular problems, and it makes the agent to seek information broadly without any distinct focus or direction. The latter is the curiosity facing to the specific problems. It is known that the specific curiosity is excited by the problem which has moderate difficulty. In other words, human is not curious about the problems that are too easy or too difficult to learn. It is expected that appropriate switching of diversive and specific curiosities improves an effectiveness and adaptability of the learning. There are some researches that apply specific curiosity to Q-learning so far[4,5]. However, the research regarded both of curiosity does not exist. We design the model that uses two types of curiosity.

3 Agent Model

Fig. 1 shows the environment settings considered in this article. Two targets move on the field according to their movement rules. A task of the agent is to learn the movement rules of the targets. The agent captures the target using sensor (gray areas in Fig. 1). The agent makes 90-degrees turns or takes a step forward at each time step, on the other hand, the targets move to either of four directions at each time step (up, down, left or right). Before the agent finds targets, the specific curiosity does not rise, thus the agent moves around the field to find targets according to the diversive curiosity. We define the diversive curiosity as a

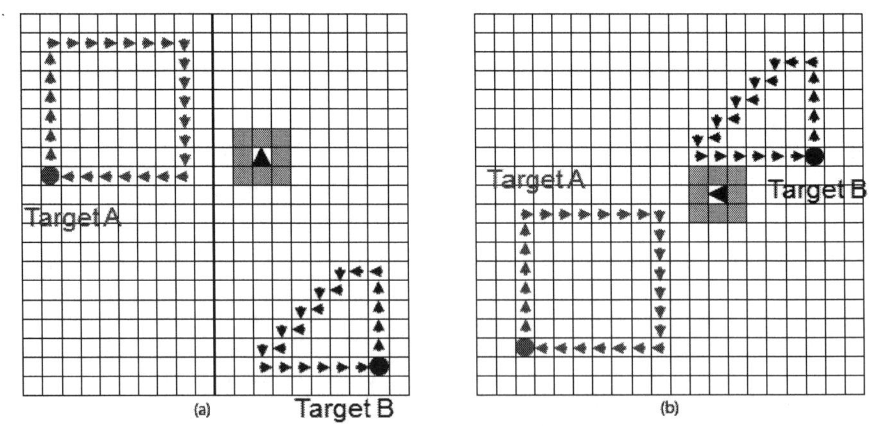

Fig. 1. Experimental environment. A task of the agent is to learn the movement rule of each target. (a)Initial moving rule of each target. (b)Moving rules after changes.

change in the degree of interest, relative to the number of times the agent visited the area. The diversive curiosity for lately visited area is decreased, and the ones for other areas are increased. We also define the specific curiosity as strength of interest to the target. When the agent captures a target, the specific curiosity about the target increases. Therefore the agent tries learning the movement rule of the target. If the agent succeeds in learning the rule, the curiosity increases according to the updating degree. On the other hand, the curiosity decreases if the agent loses the target or the finishes the learning. When the curiosity is less than a threshold *Th*, the agent moves around the field again according to the diversive curiosity. The threshold *Th* presents the ease of getting bored. Small threshold means that the agent tries learning the target persistently even if he/she sometimes loses the target. On the other hand, large threshold means that the agent gets tired of learning easily.

4 Experiment and Results

We have done some experiments to compare the proposed strategy with the traditional strategy. The experiments were done using the environment shown in Fig. 1.The specific curiosity about the target is higher than the threshold *Th*, the agent chases the target in both of the traditional and proposed strategies. Meanwhile, the curiosity is lower than *Th*, the agent moves around on the field based on random search or diversive curiosity in the traditional or proposed strategy, respectively. The total time step of the simulation is 4,000, and the moving rule of targets changes from that of Fig. 1(a) to that of Fig. 2(b) at 2,000 steps. At the beginning, we investigated the threshold *Th* with the highest effect. The threshold *Th* is varied from 0.0 to 0.9 for each strategy. For each threshold value, 10 times simulations are carried out with randomly setting the agent's initial position and direction. Fig. 2 presents the results of each experiment. The learning error becomes 0 if the agent completely learns the movement rules of two targets. Fig. 2 (a) and (c) are error corresponding to *Th* where the agent decides its action based on traditional strategy at 2,000 and 4,000 steps, respectively. Fig. 2 (b) and (d) are error corresponding to *Th* where the agent acts based on the proposed strategy at 2,000 and 4,000 steps, respectively. From Fig. 2, it is shown that small and large thresholds are not appropriate. As mentioned above, small and large threshold mean insistent and getting tired easily, respectively, thus the agent can not learn well movement rule of both target. Sometimes the agent learns one target's movement well and does not learn other target. Sometimes the agent slightly learns both targets' movement.

Next, we compares the traditional strategy (*Th*=0.4) with the proposed strategy (*Th*=0.1) on the averages of the learning error. Fig. 3 shows the comparison of learning errors of two strategies. As shown in Fig. 3, the agent quickly found the targets after movement change of the targets by using diversive curiosity (over 2,000 steps in Fig. 3). It means that the adaptability to change of environment is improved by employing the concept of diversive curiosity.

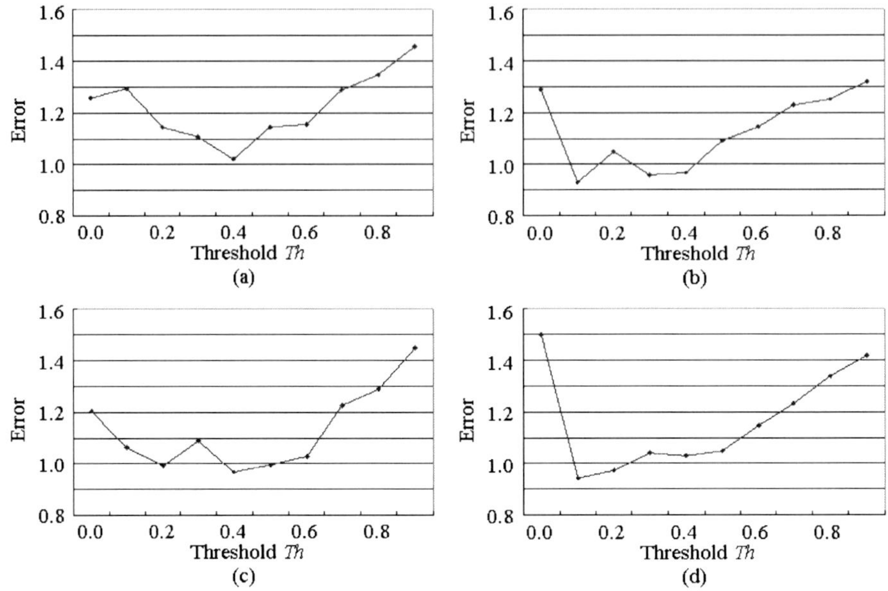

Fig. 2. Error corresponding to the threshold Th. (a)Error in traditional strategy at t2,000 steps. (b) Error in the proposed method at 2,000 steps. (c)Error in traditional strategy at 4,000 steps. (d) Error in the proposed method at 4,000 steps.

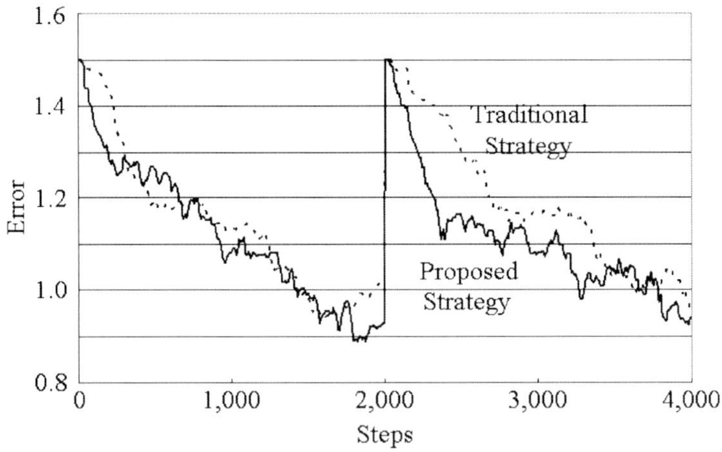

Fig. 3. Comparison of the proposed strategy against ε -greedy strategy

5 Conclusions

We proposed the action decision model based on both specific and diversive curiosities. In order to investigate the effect of curiosity, we verified the control timing of switching two curiosity modes, compared the proposed strategy with the ε-greedy strategy. As the results, the agent who cannot get tired easily can learn efficiently. In addition, diversive curiosity is effective for stable learning and adaptation to new environment.

We should be applied the proposed model to real world problems by polishing up the learning unit.

Acknowledgements

This work was supported by a 21st Century Center of Excellence Program, "World of Brain Computing Interwoven out of Animals and Robots (Pl: T. Yamakawa)" (center #J19) granted in 2003 to Kyushu Institute of Technology by MEXT of Japan. We express our sincere appreciations to S. Aou, A. Hanazawa, H. Nakagawa, T. Toyoshima and M. Mizuno (Professors of KIT LSSE) for their enthusiastic discussion about this research.

References

1. Deci, E., Ryan, R.: Intrinsic Motivation and Self-Determination in Human Behavior. Plenum Press (1985)
2. Berlyne, D.: Conflict, Arousal and Curiosity. McGraw-Hill, New York (1960)
3. Keller, H., Schneider, K., Henderson, B.: Curiosity and Exploration. Springer, Heidelberg (1994)
4. Schmidhuber, J., Munich, T.U.: Curious Model-Building Control Systems. In: Proceedings of International Joint Conference on Neural Netwarks, vol. 2, pp. 1458–1463 (1991)
5. Oudeyer, P.-Y., Kaplan, F., Hafner, V.V.: Intrinsic Motivation Systems for Autonomous Mental Development. IEEE Transactions on Evolutionary Computation 2(2), 265–287 (2007)

Author Index